EVOLUTION

EVOLUTION

By Theodore H. Eaton, Jr.
THE UNIVERSITY OF KANSAS

W · W · NORTON & COMPANY · INC · NEW YORK

SBN 393 09921 0

Library of Congress Catalog Card No. 66–11784

Designed by Nancy Dale Muldoon
Printed in the United States of America

2 3 4 5 6 7 8 9 0

CONTENTS

PREFACE

Evolutionary biology is as wide as the world of animals and plants, past and present. There is little likelihood that our ideas about it will ever become stereotyped or that our knowledge will be complete. Recent years have seen an immense expansion of evolutionary studies to many areas that were scarcely touched before. Naturally, then, among biologists of diverse backgrounds and habits of thought there may be some differing views, cross-currents of interest, and a certain amount of turmoil caused by impact among various new "waves of the future." This is not a serious difficulty and there is no harm in it, but a few who enjoy riding the crest of a particular wave may proclaim that the rest of the sea is flat and there is only one wave of the future. Some of their young students may believe them and not bother to look for themselves.

This book is meant to reflect a wide range of evolutionary principles and to offer examples showing how they work. But because the principles pervade large areas of biology, one might easily write another book discussing them in a different way, using many other examples, so that there would be almost no duplication. The chapters are intended to help the student develop a large number of ideas or centers of interest of his own concerning evolution. The first eleven regard evolution as a process affecting contemporary organisms, and the remaining six deal largely with what has happened in the past, and how we know about it.

The first chapter shows how evolutionary ideas became possible and, eventually, inevitable with the growth of biological knowledge. It was not written because a textbook ought to have a history at the beginning, but because this particular sequence is of great interest to biologists, including the writer. The present state of evolutionary theory is much more readily appreciated when we know something of its growth.

The diversity of life is very great, but it is a diversity among more or less discrete entities that can be distinguished from one another, as we see in the world around us. It is not a diversity of completely intermingled, blended characteristics. Therefore we need some way to recognize and designate these almost innumerable entities. They are the individuals, the populations, the species, and (at our convenience) the more inclusive categories we may construct. The only direct, complete continuity we discover among such separate units or groups is that of their common sources in the past. Individuals have a direct genetic continuity with their parents, grandparents, and more distant ancestors; many now living may have been born from a few. (That is why we are on the point of a "population explosion.") Species, in those cases where we find evidence of their past, have a similar direct continuity with other ancestral sources, and often many branches have arisen from one stem. Great numbers of such branches have vanished after persisting for ages and have left no descendants. A chapter on the meaning of species, both as units of classification and as biological entities, gives a foundation for later use of these concepts.

The next two chapters concern reproduction and heredity. The latter emphasizes especially the relation between heredity and environment in individuals, the sources of genetic variability, meaning of mutation, and our new knowledge of what actually is transmitted in heredity. The present understanding of populations as major units of evolution, rather than individuals or species, is discussed, and followed by a chapter on the making of species ("speciation"). This is exemplified by the problem of "species flocks" in some of the largest and oldest lakes in the world, and by the question of why there are at least a million species of insects in the world. Then the essential development of reproductive isolation is treated in a manner somewhat different from the customary emphasis on "isolating mechanisms."

Two following chapters concern adaptation and its meanings, illustrated by many examples. Several current concepts of adaptation and natural selection are examined critically, for I have taken an unorthodox view of group selection, genetic assimilation, and what is called evolutionary plasticity. In a chapter on the relationship of ontogeny to evolution the old idea of recapitulation is shown to have little use today, but, on the other hand, a change in developmental rates of certain parts of organisms, as seen in neoteny, is widespread and important as a source of adaptive novelties. Animal behavior, not generally discussed in textbooks of evolution, is given an extended chapter, and so is the distribution of organisms, both terrestrial and marine; this includes both present and past distribution in certain cases. Continental drifting, a recent revival of an old theory, is now making considerable impact in geology and is discussed here.

The geological record and the ways in which the fossil evidence is brought into the study of evolution are considered in detail, followed by a chapter on the origin of life in terms of the ancient physical and chemical environment, essential conditions for existence of life, and the beginning of biological evolution. A chapter summarizes the history of vertebrate animals, and another presents the information we have on evolution of two particular groups, one (ostariophysan fishes) without a significant record, the other (horses) in which the whole evolutionary story is shown by fossils. The last chapter, on Primates and the origin of man, is written from a zoological viewpoint, not dwelling on social or psychological hypotheses, for a student needs to know first something about what actually happened.

It has been, at times, a disadvantage but more often a help to have an almost ubiquitous interest in things biological, and the central theme of evolution has served me, as many others, for a unifying principle in what otherwise might seem chaotic views of the world. To a student who comes into an evolution course with perhaps a few quick dips into general biology, zoology and genetics as a preliminary, the situation is otherwise; he is given a framework of ideas and invited to hang his present and future learning upon it. But if from the start he realizes that the only source of authority is direct experience, whether by experiment or observation, tested and criticized until it meets general agreement among those who have done it themselves, and that he also can be in a position to obtain and interpret evidence himself, then he may find that the natural world gradually fits together into an understandable whole in his mind.

I acknowledge gratefully many discussions with colleagues at the University of Kansas and elsewhere on questions that arose during this prolonged task of writing, and especially the careful, thoughtful criticism of the manuscript by Prof. George F. Edmunds of the University of Utah, and the patient editorial understanding of Mr. Kenneth B. Demaree. Much of whatever value the book may have is due to three talented artists, Miss Kay Reinhardt, Miss Jeanne Esch, and Mr. Merton C. Bowman, who made the drawings.

THEODORE H. EATON, JR.

Lawrence, Kansas
September 1969

EVOLUTION

1 · DEVELOPMENT OF THE IDEA OF EVOLUTION

Change characterizes all living things, even dormant seeds. Life, during hundreds of millions of years, has produced an almost infinite variety of forms and activities. Some actions appear to be random, as the nocturnal movements of mice in a field, others rhythmic, as the heartbeat or the annual renewal of plants and animals with recurring seasons. Eggs hatch, young are born, and individuals repeat so closely the manner of development of their immediate ancestors that one may be tempted to think there is nothing new under the sun. Yet despite this conservatism, individuals always differ.

Beyond the cycles, wavelike rhythms, and random variations, an additional element of change appears as a characteristic of life. This is progressive or directional, in the sense that a population accumulates, over a great number of generations, some features that it did not have before and loses others that it formerly possessed. And *because* this happens, the likelihood of producing again the same combination of characteristics that it had long ago steadily decreases. The name for this cumulative change is evolution. It affects all known lineages of plants and animals, including man.

The fact of evolution has not been controversial among biologists during the present century at least, nor have serious scholars found need to question it in that time. Evolution as a real phenomenon is not a subject for debate, except occasionally among people who are either uninformed or for some reason anxious to impose dogma in the place of scientific learning. The study of evolution encompasses nearly all of biology, and the details of its processes, although extensively known, will draw the curiosity of investigators as long as scientific work is done.

The virtual absence of evolutionary ideas prior to the eighteenth century was not due to a lack of education, nor to the inability of medieval and ancient men to reason, nor even primarily to the weight of theological authority. It was due in great measure to the limited experience of individuals with the natural world, and a consequent inability to express or transmit the little experience they had or to make use of the knowledge of others. There was folklore, but those who heard it and passed it on could not easily distinguish the truth from the fiction that it contained. The tales of travelers such as Marco Polo, who went to unknown parts of the world and returned years later, might be believed when they were false and derided when they were true. The marvelous adventures of Sindbad and of Gulliver were likely to receive wider attention than an objective account of real experience.

But it was rarely that anyone traveled to a part of the earth conspicuously different from his birthplace. Change, whether at home or by geographical contrast, was far less obvious during a man's lifetime than it is today. For ages men lived near if not in their fathers' and grandfathers' homes and expected nothing different for coming generations. The cycle of the seasons was repeated endlessly, the stars and planets continued on their rounds, the familiar hills were still climbed, the same streams flowed, and even the well-trodden roads that were built before the memory of one generation could be expected to serve men not yet born. Why, then, should anyone doubt the permanence of all major features of the natural world?

Among the animals around him the countryman recognized few categories: the beasts and birds (certain kinds of each of these were known by local names), serpents (rarely could more than two or three names be applied in this category), frogs (green, brown, big, small), turtles, fishes (this category might include a dozen or more kinds, because it was sport to catch them and some were good to eat), the insects (crawling, flying, sometimes stinging, but mostly nameless vermin), and so on. In any particular area, neither ambition nor curiosity impelled men to more than a casual acquaintance with an infinitesimal fraction of the species that could have been discovered. As for creatures of foreign lands, the mythical unicorn and basilisk were as well known and thoroughly credited by most Europeans of the seventeenth century as the elephant or the rhinoceros. It was impossible, even for men who were literate and educated, to conceive of the existence of several hundred thousand recognizable species of living beings.

Apparently those who thought seriously about nature did not regard species as objective realities in the sense that we do today. It did not occur to them to suppose that species were natural populations whose integrity was preserved by reproductive barriers that prevented intermingling of their characters with those of others. It was considered entirely possible that animals and plants could originate spontaneously or by the abrupt transformation of other kinds, and yet no one put these questions to rigorous tests. Occasionally, even today, one meets a biologist who expresses some doubt about the reality of species in nature, thinking that if we had all the individuals of a genus from all the parts of their range, we might possibly find that they intermingled with each other to the extent that our supposed species disappeared. It may still be possible to adhere to such a view in a few instances—when a man's experience has been limited, for example, to Protozoa and does not include a familiarity with taxonomy of higher animals and plants—but this notion cannot be supported today.

If, in the seventeenth century, a person asked questions about the natural world, prompt and sufficient answers were available, explaining the cause (an act of creation) and the purpose (for the use, enjoyment, and enlightenment of man, the crown of creation). To this understanding it was proper that a good man might devote his life, but in practice such a life might be regarded as of little use to others, and few were drawn to it.

One of the few was the Englishman John Ray (1627–1705), a contemporary of Isaac Newton, Christopher Wren, and Robert Boyle. After traveling on the continent, visiting the naturalists of several countries, Ray settled down at home and wrote a series of books on plants, fishes, quadrupeds and serpents, reflecting the knowledge of his time. This required little investigation of the sort that we would call scientific. The teachings of his work concerned the ways in which the living creation could serve man and the wisdom of the creator in equipping animals and plants to live as they do.

Similarly, it was sufficient to Boyle and to Newton that the laws describing the properties and motions of inanimate nature be examined, so that man might know them, marvel at them, and use them, but there was no question whether nature itself might have changed its activities in the course of time. Natural laws and physical properties, they supposed, were all created at the same instant; man had only to search them out. Yet the beautiful mathematical demonstrations of physical relationships, which appeared to explain the major facts of nature, disturbed a few minds, for they suggested that nature might not have been created for man alone, that it could get along quite well without him, and perhaps might not require the immediate control of all things by an ever-watching, all-knowing creator. No one realized, however, that the processes at work in the world could slowly bring about a different kind of world.

Galileo (1564–1642) showed the character of the movements and changes of celestial bodies but was forced in his old age to deny what he had seen. More than a century later Herschel (1738–1822) and Laplace (1749–1827) finally demonstrated that there were actually no fixed points or motionless objects in the universe. Meanwhile a little attention began to be given to features of the earth itself. That fossils were the actual remnants and imprints of real animals and plants formerly living in the world had been understood by the ancient Greeks but in the meantime forgotten until rediscovered by Leonardo da Vinci (1452–1519). That mountains bore indications within them of the way in which they might have been formed was gradually appreciated by a few men (Hooke, 1635–1703; Steno, 1638–1686; and Arduino, 1714–1795), and the implications for the earth as a whole were dimly seen. Some mountains, for instance, were volcanoes, perhaps no longer active and no longer retaining their original shape, yet built up originally by the flow of lava and the eruption of volcanic ash; others clearly indicated a folding of beds that had formerly been horizontal, or a lifting up, tilting, and breaking apart of strata that were previously undisturbed.

The major restriction on the development of this kind of knowledge was the rigid, enforced belief that any changes that had taken place did so within the 6000 years since the supposed date of creation. Thus it was not possible to understand the accumulation of huge quantities of sediment at rates now observable; the biblical legend of a universal flood was used to explain both the deposition and erosion of the earth's surface features. Furthermore, as the world was presumably perfect at the start, we could look forward (barring additional acts of creation) only to the decline and ultimate ruin of all things, at no very remote date.

Buffon, the encyclopedic naturalist of the eighteenth century (1707–1788), sought the reasons for natural phenomena in a spirit more modern than that of his contemporaries. He felt compelled to extend the lifetime of the earth to six epochs totaling perhaps 70,000 years, and yet was not satisfied with description only in his account of living things but inquired into their similarities, differences, distribution, possible modifications by or for their environment, and their variations. But he inferred that as variability was random in character, any species must therefore be deprived of the perfection with which it was originally endowed; the result would be slow decline and final extinction.

In preevolutionary days some of the more imaginative thinkers attempted to explain in a half-natural, half-mystical way common patterns or forms of living things, among which it often seemed as if a few themes prevailed but with endless variations. For example, Johann Wolfgang von Goethe (1749–1832), perhaps the most gifted of all men of his time, saw the forms of flowers, leaves, bracts, thorns, and even of tendrils in climbing plants as variants of an ideal or perfect leaf, brought to realization by nature in a mystical way (*Metamorphosis of Plants,* 1794). This "natural philosophy," with its tinge of romantic poetry, expressed remarkable insight but was not sufficiently critical or thorough to serve as a strong foundation for biology.

We cannot easily distinguish among the various interests of these early scientific men, nor even between scientists and philosophers, for they themselves did not make such distinctions. John Hutton (1726–1797) could, however, be called a geologist almost in the modern sense. In his *Theory of the Earth,* first published in 1788 and again, enlarged, in 1795, he interpreted the past by his studies of present processes and set no limits of time for either the beginning or the end of the world. Few were prepared to accept these views in his day, and controversy over Hutton's work was prolonged, especially on the very difficult matter of man's creation; this, if nothing else, must be held within the limits set by theologians if one wished to gain any hearing at all. The idea of natural processes acting continuously throughout a vast expanse of time, without interruption and without perceptible beginning or end, was one of the essential premises for our later understanding of evolution, but many years passed before it won general agreement.

William Smith (1769–1839), a practical geologist, found that stratified rocks could be traced and identified across large areas by means of the fossils they contained. It was customary in the eighteenth century to regard fossils in general as the "wrecks of a former creation," that is, of antediluvian life, prior to the flood of Noah. In France, Georges Cuvier (1769–1832) and his associate Alexandre Brongniart saw that the kinds of fossils changed as one went from lower to higher strata, and also that fossil mammoths and rhinoceroses were obviously like (although not precisely the same as) living elephants and rhinoceroses, but Cuvier preferred to imagine a series of prehistoric epochs ended by worldwide catastrophes to account for these changes rather than supposing a continuity of life, with gradual evolution.

Charles Lyell (1797–1875), in his *Principles of Geology* (1830–1833), was finally able to bring together a quantity of geological evidence great enough to settle, once and for all, the question of uniformitarianism versus catastrophism in the past history of the earth. Whereas earlier students of geology had supposed that all crystalline or amorphous rocks must be "primary," Lyell was able to show that in the vicinity of volcanic activity, past or present, there was disturbance and a metamorphosis of some sedimentary beds to a crystalline condition, the fossils contained in them being destroyed in the process. This recognition of metamorphic rocks as a secondary category clarified many problems of interpretation. Lyell also showed, to his own astonishment, that fossil shells in limestone beds several hundred feet thick on the coast of Sicily were of species now living in the Mediterranean Sea. He was able to demonstrate the gradual replacement of extinct species in beds of Tertiary age by other species that survive today. His familiarity with fossils and with the natural history of living marine and freshwater animals made it possible for him to interpret past environments from the fossils in sedimentary rocks.

At about this time discoveries of prehistoric "monsters" were being made in America. It is a remarkable fact that early in the eighteenth century the teeth of mammoths were recognized by Negro slaves on southern plantations as belonging to elephants, when the same teeth were attributed to human giants by such educated, literary men as Cotton

Mather. Later, Benjamin Franklin and Thomas Jefferson, among others, became reasonably sure of the correct identification of mammoth bones and teeth. Jefferson was not persuaded that the animals actually were extinct.

Meanwhile, a French zoologist, Jean Baptiste de Lamarck (1744–1829), presented an essentially correct picture of the way in which different species of animals could be related to each other by descent along branching lines of a "family tree" in his *Philosophie Zoologique* (1809). This was an actual theory of evolution, for he thought that animals became progressively adapted to particular ways of life, into which they fitted more and more perfectly, and he offered an explanation of the means by which this was done. He supposed that the *needs* of individuals, as reflected in their activities, brought about internal changes that helped to improve particular parts or that, by disuse, weakened and gradually reduced these parts, such effects being inherited. The idea was entirely reasonable in his day, for it seemed to accord with everyday experience and it clearly accounted for adaptation as a characteristic of life. But Lamarck's rival, Cuvier, could see no evidence of this kind of change in his wide zoological experience and would not grant the possibility that species, having been created, could change without a new creation. His opinion prevailed at the time. Moreover, no one knew the mechanism of heredity; the evidence against the inheritance of characteristics acquired during a lifetime did not become clear to biologists until the twentieth century. The Lamarckian explanation of adaptation is not now accepted.

Now, having traced the larger steps in thinking through which a theory of evolution became possible, we must take note of a signpost set up by Linnaeus (1707–1778) in his *Systema Naturae* (1735; the tenth edition, to which reference is usually made today, was published in 1758). His conception of the species and genus was essential for the growth of biology, not only because two names for each kind of animal and plant were more convenient than descriptive Latin phrases, but because the names themselves were a means of indicating natural resemblances or differences among the "kinds." Indeed, for a clear understanding of modern studies of evolution it is necessary to appreciate the concepts of species and genus as well as of the higher categories—family, order, class, and phylum. This is the subject of Chapter 2.

Linnaeus, of course, was not an evolutionist, nor was he greatly concerned with explanations of nature, but he devoted his life industriously to arranging the kinds of animals and plants known to him in a convenient system. This system he took to be actual and fixed; to discover it and use it was the major task of natural history. He broke with tradition in placing man, "the knowing one," *Homo sapiens*, in the animal kingdom, the class Mammalia and the order Primates. Of the many categories employed today in classification, it is interesting that Linnaeus used only the kingdom, class, order, genus, and species, but other naturalists soon found it advantageous to divide the animal and plant kingdoms into phyla and to divide orders into families. In fact, it has frequently happened that a genus, as described by Linnaeus, represented what is now a family or even higher category, because of the later discovery of numerous and diverse species that were unknown to him.

Incidentally, when Linnaeus established successfully his idea of species, as well as the terminology used to express it, he created a dilemma that gave the early naturalists additional trouble before they could think their way through to a concept of evolution. This dilemma was that a species is real, is objective, and retains its identity in nature instead of shifting haphazardly or spontaneously changing its characteristics. On the other hand, if he were to accept the origin of species from others, then he could hardly agree with the Linnaean idea of what a species is. The solution of this problem came slowly during the Darwinian period in the nineteenth century. It is, of course, that species are composed of populations that, like flowing streams, retain their integrity while undergoing gradual change.

At this point we can see in retrospect how an intellectual climate had developed in which a theory of evolution became possible, a theory that envisioned the origin of existing species from earlier and different ones and the descent of varied plant and animal forms from a common ancestry in the remote past (see Greene, 1959, 1961). The development of this climate came through gradual realization of the following facts: Given ample time, there are changes in all things on earth and in the outer universe; the earth carries its own record of a sequence of events far too vast to have been accomplished within a few thousand years; processes now at work have also worked without significant interruption in the past; many kinds of animals have disappeared that formerly existed, and conversely many have appeared that were not present in the earliest times; living beings are endlessly diverse and varied; it is possible to recognize, name, and systematically arrange many thousands of species of plants and animals, including man; the real meaning of such a system is that it expresses natural relationships, and the relationship itself is a result of descent, with divergence, from common ancestors; this descent is accompanied by cumulative and generally adaptive changes. These facts could be recognized early in the nineteenth century but probably not by more than two or three men.

The story now becomes much more complex and moves with increasing rapidity toward the tremendous expansion of biology in the twentieth century. But it is well to remember that in the second half of the twentieth century, and in the most advanced countries in the world, some of the major concepts that became evident in the eighteenth century have not yet found their way into the minds of all who can read and write. (As a single instance of this, note the fantastic revival of a notion of catastrophism in a book by Velikovsky, *Worlds in Collision,* highly popular a few years ago.)

Charles Lyell, whose conception of uniformity in natural processes has already been mentioned, also discussed in his *Principles of Geology* (1830–1833) the problem of species. He supposed that species of animals and plants were, in general, created in and fitted for the places they occupy, but he saw that extinction took place frequently and that changes in distribution were likely to occur. Environmental changes would affect some kinds of plants and animals favorably, allowing a spreading and increase, while other kinds suffered reduction, a harsher struggle to survive, or failure. He conceded with some hesitation that new species might originate as older ones were lost but could not accept Lamarck's explanation of the causes of their origin; he remained noncommittal as to the fact or method of evolution.

Charles Darwin (1809–1882), in his early twenties, was engaged as a naturalist on the voyage of the British naval vessel Beagle around the world. The ship cruised down the eastern shore of South America, stopping for long intervals in several places, including Patagonia and Tierra del Fuego, then passing around the lower end of South America and up the west coast. It stopped for another interval among the Galápagos Islands on the equator and later sailed westward across the Pacific and on around the world. While Darwin was in South America and the wonderful fauna and flora, the geology, and prehistoric life of that land were being revealed to him, he had the opportunity to read Lyell's work on geology. This started in his mind a train of thought that led to his own great accumulation of evidence bearing on the origin of species. He collected extensively in Patagonia, Tierra del Fuego, and the Galápagos Islands, some 600 miles off the west coast. On these semiarid volcanic islands he found that different species of giant tortoises lived on different islands, yet were obviously closely related to one another and might have been derived from one ancestral stock that originally populated the islands from the mainland. The same was true of a group of birds, the ground finches. But Darwin was, throughout his life, exceedingly cautious about publishing conclusions from his evidence. Although the number and size of his publications are considerable, all were the results of long deliberation.

After the voyage ended in 1836, Darwin began accumulating information on the variation of animals and plants in domestication and on the manner in which selective breeding changed their characteristics. He followed the advances of biology, agriculture, and anthropology, carried on an immense correspondence with botanists and zoologists, dealt with the classification and natural history of the barnacles, and made extensive studies of the adaptations for pollination in plants, of the role played by earthworms in soil fertility, of geographical distribution and dispersal, of the expression of emotions by animals and man, and eventually of the descent of man. But the central theme of his activity was a concept that he first glimpsed, apparently, in 1838, after reading an essay on population by Malthus (1798),—that a process of selection in nature could accomplish changes in wild species even more profound and universal than were achieved by man in his artificial selection of domestic stock. The idea simmered in his mind for four years before he ventured to put it in writing. Not until 1844 did he develop his theory in an essay, but not for publication; he showed it to no one but Lyell, because he was still remarkably diffident about expressing his views. In 1856 he finally began putting together, out of a mountain of data, a book on the origin of species.

Meanwhile, Alfred Russel Wallace (1823–1913), an English naturalist who spent many years collecting and exploring in South America and the East Indies and who contributed greatly to knowledge of animal geography and adaptations, independently thought of natural selection as a method for bringing about the origin of new species and in 1858 sent to Darwin a brief essay on this topic. As a result of pressure from close friends, Darwin was induced then to make a statement of his own view, with those of Wallace, and to publish them jointly in the Journal of the Linnaean Society. Under continued urging, Darwin then wrote what he called an "abstract" of his intended work, and this was published in 1859 with the title, *On the Origin of Species by Means of Natural Selection.*

This great book appeared precisely at the time when an effective explanation for evolutionary changes was being sought and when abundant facts, thoroughly integrated and long considered, were essential. Darwin accepted in part Lamarck's theory of the way in which changes were caused, but considered natural selection to be more effective.

According to the theory of natural selection, variations (some of them heritable) occur in all populations of living things, and because some of these variations result in greater, others in less, probability of survival, the former are likely to be sustained in coming generations while the latter either decline or are eliminated at once. The fact that reproduction always brings more individuals into being than are able to reach reproductive maturity, and that the environment never permits survival of all offspring of a species in a generation, means that an advantage may be given to individuals having favorable adaptive characteristics. There is a pressure tending against survival to maturity or, in opposite terms, a "struggle" in which some characteristics make survival more probable than do others. This does not mean, however, that Darwin thought natural selection worked solely by ruthless "survival of the fittest" or that the "struggle for existence" constituted an unending battle. Selection could also work effectively by such means as the production of nine offspring by one pair of parents as compared with six or two by another. A final element in the theory, not fully appreciated by Darwin, is that the divergence of species from a common ancestral stock is brought about by some form of segregation in the original stock, so that the long-term effects of selection in one part of the population are different from those in other parts and isolation prevents a remingling of the hereditary traits thus changed. Without barriers to interbreeding, races or species could not become distinct, but with such barriers, they would eventually, inevitably, do so.

With the powerful support of T. H. Huxley (1825–1895), the idea became widely established among biologists that evolution had occurred and that it could be explained by processes observable in nature. New lines of investigation developed rapidly in the latter half of the nineteenth century, only two or three of which are especially pertinent here.

The English anatomist Richard Owen (1804–1892) and the German Karl Gegenbaur (1826–1903) illustrate, between them, the transition from preevolutionary to evolutionary interpretations of structural patterns in animals. Owen did not, until late in his life, accept evolution as a fact, but in his extensive studies of anatomy he recognized common patterns for such structures as the skeleton of the limbs or the vertebrae of different animals, and in the concept of homology he expressed something like the older belief in variations from an ideal type. Gegenbaur, to be sure, pursued structural similarities among animals in much the same way, but with a clearer conception of evolutionary descent. Having little detailed evidence from the fossil record as to the actual course followed in the evolution of the skeleton, for instance, he nevertheless traced the derivation of limbs, skull, gills, and other parts to their "archetypes," largely found in the structure of animals still living. Reliance upon this approach persists even today in the teaching of some biologists, but as a result of the rapid growth of paleontology it is seen that the fossil record, when sufficiently complete, may show that an ancestor was not like anything now living. The "ideal" vanishes when the real is discovered; the "archetype" is replaced by the ancestor, which may be very different.

Much the same comment should be made on our newer understanding of species and other systematic categories. The Linnaean species, with its selected "type specimen" and the rigid formalized rules under which it obtains recognition, is no longer the heart of systematic biology. Rules are indeed essential to avoid utter chaos in classification, and type specimens are valuable as objects of reference, bearing the name given by the author of the species, but the "type" does not mean a "typical" specimen, and it has no more real value in describing the species than any other specimen of the same population. The *species* is not a name, a concept, a museum series, or a diagnostic set of characters; it is one or more populations of animals (or plants) capable of fertile interbreeding among themselves, or, in the case of nonsexual species, the offspring or parents of other individuals in the same population, whose characteristics intergrade with their own.

We recognize, below the rank of species, subspecies, which are usually geographical races, and give them names; interbreeding usually takes place between members of different subspecies if their ranges meet. As a result of thorough study in familiar groups of organisms it has become evident that species are often divisible into subspecies, that species now existing are usually but not always recognizable as such by their appearance, and that the definition or description of every species breaks down when we are able to trace its lineage back in time, for we can observe differences in the characters of ancestral populations beyond the limits of those existing at this particular moment.

Another line of biological investigation, which received its first strong impetus about 100 years ago and has contributed increasingly to the understanding of evolution since then, is paleontology. Although invertebrate fossils greatly outnumber those of vertebrates and are of more practical value to the geologist, the study of the higher vertebrates (reptiles and mammals especially) has given biologists at large their most vivid conception of evolution as a history. Beginning with discoveries of certain spectacular fossil animals that aroused curiosity but were not well understood, vertebrate paleontology has grown through a series of stages, roughly these:

1. The rapid increase in number of genera, families, and orders, with accompanying shifts in classification, as previously unknown creatures were collected and described; all too often the descriptions were inadequate, leading to much revision later.
2. The recognition of many phylogenetic lines, some of these both extensive and detailed, as in the horse family.
3. The establishment of ancestral connections among the larger categories; for example, mammals were shown to have descended from an order of mammal-like reptiles instead of from amphibians, as was suggested by Huxley in the mid-nineteenth century.
4. The precise determination of sequences in sedimentary deposits, recognition of successive faunas and floras in past geologic ages, and interpretation of environmental conditions in the past. Among the eminent workers in vertebrate paleontology we may note especially Edward D. Cope (1840–1897), Othniel C. Marsh (1831–1899), H. F. Osborn (1857–1935), W. D. Matthew, W. K. Gregory, Alfred S. Romer, and George G. Simpson in America; A. Smith Woodward (1864–1944) and D. M. S. Watson in England; and Robert Broom (1866–1951) in South Africa.

The last field of study to be mentioned here as a foundation of modern evolutionary theory is genetics. Although a knowledge of the mechanism of heredity does not give us an understanding of the course of evolution or of all the circumstances that influence it, the evolutionary process is, almost by definition, a change of inherited characteristics. The means by which these originate and are transmitted, generation after generation, in every population of organisms, must therefore be a major premise in the solution of any evolutionary problem. But this particular premise was not yet available to Lamarck, Cuvier, Darwin, or, in fact, any student of evolution before the year 1900.

In that year the attention of biologists was first drawn to some experiments carried on in Austria by Gregor Mendel (1822–1884) and described in a paper published in 1866. By crossing different strains of peas and keeping account of the number and characteristics of all offspring of each cross, he demonstrated that (1) the factors given to the offspring by the parents do not "mix" but are segregated, and (2) if more than one pair of contrasting characters are considered in the same cross, the factors responsible for these are inherited independently.

The meaning of the observations became clear in the light of August Weismann's (1834–1914) theory of the "the continuity of the germ-plasm" (1892). The protoplasmic material responsible for heredity maintains its identity or, as we say now, replicates itself, through the life of each individual as well as from parents to offspring. Further, Mendel's observations accorded precisely with discoveries made between 1875 and 1887 by Oskar Hertwig and Edward van Beneden, of the changes in the nucleus of the egg and of the sperm cell at fertilization. It was possible then to see that the segregated factors of Mendel behave as they do because they are associated with chromosomes, the self-maintaining, independently segregating bodies that are present in the nuclei of cells.

One of the "discoverers" of Mendel's work, the Dutch botanist Hugo de Vries (1848–1935), believed that he had found, in 1900, the method by which new species originate—not in natural selection, but in the rare occurrence of spontaneous variations (*mutations*) that could be inherited. This appeared so simple and obvious from his example of a new kind of evening primrose appearing suddenly in a greenhouse, that for the next thirty years many biologists supposed that "Darwinism" was now obsolete and that evolution was no longer an issue worthy of serious consideration. Thomas Hunt Morgan (1866–1945) and his associates developed the theory of the gene by intensive experimental study of heredity in the fruit fly, *Drosophila*. The complexities of this study have been such as to absorb the energies of innumerable students during more than fifty years.

In the third and fourth decades of the twentieth century the attempt to apply genetic information to conditions existing in nature resulted in many improvements in understanding. One of these came from mathematical evidence (J. B. S. Haldane, R. A. Fisher, and Sewall Wright) that natural selection can, after all, account for genetic changes in a population, given the mutability of genes as a source of selectable variations. Another was that some problems in taxonomy, distribution, and behavior of animals, as well as in their geologic history, are clarified by a knowledge of genetics. To this might be added clear demonstrations of natural selection at work, modifying the characteristics of certain natural populations.

Perhaps the most significant development in the last twenty-five years is in the ability of students of evolution to review critically, and to summarize in a comprehensive manner, the bearing of evidence from all parts of biology upon the principle of evolution. Such work has been done repeatedly by a number of writers, especially Theodosius Dobzhansky (1951), Julian Huxley (1942), George G. Simpson (1951, 1953), and Ernst Mayr (1942, 1963). The current concept of evolution broadly resembles that of Darwin, with the addition of a hundred years of accumulated evidence on the history of life, the genetic and physical nature of living things, and the interrelationships between them and their environment.

REFERENCES

Dobzhansky, T. 1951. Genetics and the Origin of Species, 3rd ed. Columbia Univ. Press, New York.

Greene, J. C. 1959. *The Death of Adam; Evolution and Its Impact on Western Thought.* Iowa State Univ. Press, Ames.

———. 1961. *Darwin and the Modern World View.* Louisiana State Univ. Press, Baton Rouge.

Huxley, J. 1942. *Evolution: The Modern Synthesis.* Wiley, New York.

Malthus, T. R. 1798. *Essay on Population.* J. Johnson, London.

Mayr, E. 1942. *Systematics and the Origin of Species.* Columbia Univ. Press, New York.

———. 1963. *Animal Species and Their Evolution.* Harvard Univ. Press, Cambridge, Mass.

Simpson, G. G. 1951. *The Meaning of Evolution.* Yale Univ. Press, New Haven.

———. 1953. *The Major Features of Evolution.* Simon and Schuster, New York.

2 · SPECIES IN CLASSIFICATION AND IN NATURE

The title of Darwin's major work may have accustomed biologists to thinking of evolution as a process of originating new species. This, of course, is of great interest and importance and is being studied intensively at the present time, under the name of "speciation." It is but one of many results or aspects of evolutionary change that will be discussed in this book. An essential step in understanding modern evolutionary theory is, then, the consideration of *species* as they have been and are now understood, and their relation to other kinds of categories, both lower and higher. The subject is complex and further treatment in terms of genetics and evolution will come in later chapters.

There are two ways in which we may regard species and other categories of classification. One of these is that they are convenient systematic units, and classification is a complex discipline with its own necessary rules and a broad theoretical basis. Species could be recognized and arranged in a system even if we knew nothing about their biology or evolution. This taxonomic approach, incomplete though it may be, will always have its valid uses. Indeed, for the majority of species of animals and plants it is still the only approach that we have and provides all our present information about them. On the other hand, species are realities in nature. Many of them can be observed, sampled, their populations and ecology investigated, their life histories described, their genetics analyzed, and even something of their evolutionary history can be traced.

Both of these ways of looking at species are valid and necessary. They are not in conflict (although the biologists who use them may be). But they have different purposes, and they both need to be understood. Therefore this chapter considers first the nature of species in the traditional sense, as, for example, a collector sees them, and then, in more detail, the biological and evolutionary concepts of species and other categories.

SPECIES IN THE TRADITIONAL SENSE

Linnaeus, in his *Systema Naturae* (1758), crystallized the classification of animals and plants in a manner appropriate to his time. There already existed a large and cumbersome terminology in Latin, used by naturalists, in which a group of animals was designated by a name and each recognized "kind" within the group specifically indicated by descriptive adjectives or a phrase. This was on the same principle as saying "deer, white-tailed; deer, black-tailed;" and so on. Aside from his systematic cataloguing of the kinds known to him, the major innovation by Linnaeus in nomenclature was to condense and simplify names so that a single word would be the group (or genus) name and a single word following this would be the name for a specific kind (species), such as *Passer domesticus, Mus musculus,* and *Homo sapiens.* Although he and others usually tried to keep the names sufficiently descriptive to provide some meaning, description is not an essential part of the name itself, either for species or for any other category.

Species were then, and usually are today, represented in collections by specimens, which are stuffed skins, skulls, eggs, pinned insects, preserved fishes, seashells, pressed plants, and so on, according to the nature of the material collected. There were only two problems in classifying a collection: one, recognizing resemblances, the other, noticing distinctions. Specimens that resembled one another so closely as to provide a completely intergrading series not separable into groups by any distinctive features were placed in a single species, even though minor individual differences might be observed. They were treated as duplicates, like copies of a book in a library, or like a series of postage stamps in an album.

But individual animals and plants differ in some degree, and it is possible in such a series to arrange individuals so that they show a range of variation, as from a minimum to a maximum size, from a pale to a dark color, in the form of various parts, and so on. In this case variation does not imply distinctness, because, within a species, there is usually no point at which groups can be separated without intergrading. This means that any variable characteristic such as size, weight, color, or the shape of particular organs could serve as the basis for arranging specimens in a continuous series. A series based on color might go from light to dark, or bright to dull, or large spots to small spots, but there would be no point along that series at which there was a distinct break. At the same time, a series could be made from the same specimens, perhaps arranged in different order, which would extend from smallest to largest, from most hairy to least hairy, or from the smallest number of scales to the largest number of scales. And again, although it did not correspond with the series based on color, there would still be no gap separating the specimens into two or more groups.

On the other hand, the collection might contain many individuals that must be placed in different series (species) because distinctness (not intergrading) sets them apart as different "kinds." Therefore, a species was to Linnaeus a group within which individuals resembled one another closely but between which and another such group there was an unbridged gap. In the eighteenth century naturalists were already well aware that most of these species represented living populations, and that more individuals of the same kind could be found by going and collecting again in the same place as that from which previous specimens were taken. They recognized that the resemblances were due to transmission of characteristics from parents to offspring and among the members of a population, and they knew the geographical limitations of at least some species.

To Linnaeus and his contemporaries, the problem of a common origin of related species, or of the change of species in time, did not arise. Categories, names, and descriptions were intended for classifying the contents of collections and showing the fixed order of the natural world. The concept of species was based upon what could be seen, that is, on morphology or form, and it was customary to think of a species as a "type," subject to minor and unimportant variations.

To any student of animals and plants it would soon become apparent that there is a great difference in the size of the gaps between species. These gaps are sometimes small enough so that only meticulous examination of specimens will detect any constant distinction. But in certain cases a species will appear to be remote from any others known and different in almost every respect, yet it, too, must be placed in a genus. Therefore the category of *genus* is not just a group name for several closely related species but frequently designates a species that is too far separated to warrant placing any others with it. Such a genus is monotypic. The characteristics of a genus are not merely those held in common by a group of species but may also indicate an amount of difference great enough to warrant a higher rank than species, in case they occur in a single species that has no close relatives. Thus *sapiens,* for example, was the only species of *Homo* known to Linnaeus, but he recognized more than one species of *Ursus* (bears), and later other workers named more genera as well.

The expressions "specific characters" and "generic characters" thus would imply a judgment of different degrees of distinctiveness in the features used to recognize the two categories. The genus was, and still is, in a sense, artificial, being based upon the opinion of its author, and may be changed in scope as new species become known. Most of the genera named by Linnaeus have received so many additional species during the

200 years since publication of his *Systema Naturae* that they can no longer serve as useful genera, but have been elevated to families or even to orders.

In the practice of classification, these remarks concerning the genus apply also to the family, order, class, and phylum. Any species receives a place in all these categories. Any genus is placed in a family and any family in an order, and so on. Thus we have a hierarchy of at least seven ranks, including the kingdom, that are always used and are obligatory, but many more are optional, depending on the circumstances. The classification of large and complex groups such as mammals, fishes, or insects requires such ranks as superfamily, subfamily, superorder, suborder, sometimes infraorder, and others. The maximum number is more than 20, and the necessity for these is imposed by general agreement among students of particular groups. For example, the rank of subclass may be helpful in many instances but not in others; the class, however, is always used.

The modern practice of classification is *taxonomy*. Any single unit of rank, such as a subspecies, species, genus, family, and so forth, is a taxon (plural, taxa). One might consider taxonomy an art, in the sense that its practitioners create a replica of what they understand the orderly arrangement of nature to be, but it is even more clearly a science, in that the taxonomist has first to discover and interpret the evidence that he considers to indicate that natural order.

Many rules, and customs with the force of rules, are employed to maintain stability amid the frequently necessary additions and shifts in classification. The International Commission of Zoological Nomenclature deals with differences of opinion regarding the application of these rules in the taxonomy of animals. Taxonomic rules are most rigid in the case of species and genera, but the categories above the genus are subject to greater flexibility because extensive changes in their scope are necessary from time to time as the understanding of relationships develops. Only the most thoroughly studied groups, such as birds, are now in a reasonably settled state in which it is unlikely that many more new species or genera will be discovered or that present opinions of the characters of major groups are seriously wrong. But the great majority of orders and classes of animals are far from completely known, and for many our information is just beginning to accumulate. In general, the practice of taxonomy today is essentially like that of the past but greatly clarified by many kinds of information that were not available to earlier biologists.

MODERN CONCEPTS OF SPECIES

Characteristics of Species. The vast majority of species of animals and plants in the world, as far as they can be represented in collections by specimens, must be described and classified in much the same way as in the past because the taxonomist has no other information than that available to him in the collection, in publications of other taxonomists, or in the field records of the collectors. He may or may not concern himself with the meaning of species. In nearly all cases, his work is adequately done without such considerations, but an evolutionist today is greatly concerned with understanding the biological nature of species.

The concept can be expressed in many different ways according to the criteria available. This does *not* mean that there exist in nature a corresponding number of different "kinds" of species. The criteria may be reproductive methods, population structure, genetics, relationship in space or time with other species, and so on, but these criteria may not be mutually exclusive. It is increasingly clear that no single definition of species can be devised to express all the actual meanings of the word. Species are realities in nature, but the struggle to construct a universal definition defeats itself because it cannot reasonably express the facts associated with all valid concepts of species. Species are diversified phenomena to be *characterized* rather than "defined" as a whole.

One characterization is that a species consists of individuals among which the variations overlap or intergrade in such a way that no constant, clear-cut distinctions among them can be seen other than those due to age or sex. The criterion here is *morphological,* and it is satisfactory for many purposes, as the identification or description of shells, butterflies, lizards, and so on, in a collection, or the listing of a fossil "fauna," that is, the kinds of specimens occurring in a particular deposit of a certain age. But it does not serve in those cases in which the only difference that can be found to separate one species from another is in a physiological trait, behavior, breeding season, or other nonstructural feature (see sibling species, p. 22). Also there are polymorphic species in which variations of certain characters (color, shape, size) may be discontinuous, even within the same population, so that individuals apparently too different to belong to the same species may be produced by one pair of parents (see p. 28).

Another widely useful characterization is that a species is one or more populations of individuals among which breeding takes place. This is the "biological" or *genetic* definition. Naturally it is limited to species that have sexual reproduction and it is of no use for verifying taxonomic work on fossils, or in species that are inaccessible for breeding tests or known only from a few specimens. Such a genetic characterization of species is implied in the thinking of most biologists about most kinds of organisms, but as a test, or as a definition of all species, it obviously cannot be used.

For species that reproduce by budding, fission, parthenogenesis, spores, or other asexual means exclusively, the term *agamospecies* has been proposed (reproduction without gametes). This, indeed, sets a certain category apart, although it does not suggest how, or even whether, any concept of species may be applied to such organisms. Some authors, as Dobzhansky (1951), have supposed that it cannot, because there seems to be no genetic continuity within a population at a given time. This view reflects a commitment to the genetic definition, but to a student of nonsexual protozoa or parthenogenetic plant lice, there is no unusual difficulty in recognizing the species on purely morphological grounds. They evidently exist, although their reproductive nature is different from that of other species. Asexual methods of reproduction have originated several times independently among animals, but there are very few groups that are exclusively asexual. This is also true of plants.

Among organisms that reproduce asexually, an individual is descended directly from just one parent and in turn gives rise to other individuals without an exchange of genetic material but rather by fragmentation or multiplication of itself. In this case, a species can be said to consist of a number of *clones,* each clone being those individuals that are derived from one progenitor. Theoretically, it is possible that all members of an asexual species might belong to one clone. It is obvious that the problems of species formation in asexual animals and plants are quite different from those in species that reproduce sexually. For instance, the necessity for reproductive isolation is not involved because individuals are already isolated reproductively, and presumably natural selection brings about its effects in varying ecological circumstances without any need for mechanical or geographical barriers.

There are many organisms in which both sexual and asexual methods of reproduction occur, and the characteristics of a clone can be seen, therefore, temporarily, in a cycle of these alternating methods. For example, in temperate countries aphids, or plant lice, reproduce sexually in the last generation of the season. The fertilized eggs go over winter and hatch into parthenogenetic females in the spring. These may reproduce without males for fifteen or twenty generations during the summer, but males occur in the last generation in the fall. It is possible, then, for thousands of plant lice to be produced in the course of a summer from a single egg that hatched in the spring, and all, of course, carry identical genotypes if there are no mutations during that time. Among tropical aphids, the sexual generation is usually eliminated and parthenogenetic reproduction continues throughout the year. Of course, it is among bisexual organisms that the greater part of the study of species in nature has been carried on.

The criteria to characterize species mentioned so far are (1) morphological (based on visible characteristics readily used by the taxonomist), and (2) reproductive (the fact of interbreeding in nature in bisexual species and the absence of this [requiring a morphological approach] in nonsexual or uniparental species). Because these criteria cannot be fully definitive or apply in all circumstances, several descriptive terms are used for still other cases, and these must be discussed briefly.

Sibling species (Mayr, 1942, 1948, 1963) are those that resemble each other so closely as to be indistinguishable, or barely distinguishable, by any visible characters, yet which do not interbreed in nature even though they may be sympatric. Attention has been drawn to them by a behavioral, physiological, or genetic difference. In fact, the usual genetic incompatibility that is found between species has been established among these "siblings" and is ordinarily accompanied by several other differences, although these do not entail many, or perhaps any, morphological features. Sibling species are widespread among bisexual animals but are not common. They do not constitute a special phenomenon or a "kind" of species but are simply those that, having previously been confused with each other, turn out to be separate, as they would have in the first place if enough had been known about them. In most cases anatomical differences have been discovered upon close examination. Examples are the fruitflies *Drosophila pseudoobscura* and *D. persimilis*, a group of six species of *Anopheles* (mosquitoes) in Europe, and the boat-tailed grackles (large blackbirds) of the Gulf Coast (*Cassidix major* and *C. mexicana*). There is no reason to suppose that sibling species are necessarily in an early stage of evolutionary divergence, although some of them may be.

The student of fossils meets special taxonomic problems, not only practical but theoretical. His material may be fragmentary; indeed it is never complete in the sense that recent organisms are, because in general only hard parts are fossilized and these may be scattered or badly preserved. Yet they can and should be described and dealt with taxonomically, insofar as they present characters. Frequently a species or a genus is known from a jaw with a few teeth in it or even less adequate material. Some biologists question the validity of taxonomic treatment of such specimens because they suppose that a species should be characterized in full and the limits of its normal variation pointed out. But a species can also be based on the degree of difference between the characters of one or a few specimens and those of any others previously known; the nature and degree of such differences are familiar to specialists who have worked with hundreds or thousands of fossils. The idea of species, applied to fossils, then meets practical difficulties affecting the comparison or full description of the organisms concerned, but it encounters even more serious theoretical difficulties.

At any particular time, most species of animals and plants consist of (potentially) interbreeding individuals occupying particular geographical areas. In any single region there are many different species that remain different because they do not, for one reason or another, interbreed. But when the dimension of time is added, any particular species can be expected to undergo evolutionary change of its characters (phyletic evolution), such that specimens of a given age, although connected by a continuous series with others directly ancestral to them, perhaps a million or more years earlier, differ from these sufficiently to warrant recognition as another species. Simpson (1961), following Imbrie (1957), would call these *successional* or *successive* species; the terms paleospecies and chronospecies are also used. But the question is how to distinguish them, even when the information is wholly adequate to show their characters and to indicate the reality of the connecting lines. Simpson (1961) suggests: "When in a lineage the inferred ranges of observed changing characters for populations at two times do not overlap, those populations may be placed in different successive species and the division between species may be drawn approximately midway (in time) between them." Any previously named "species" in this series should be recognized as far as possible, simply because they are already a matter of record. Any gap in the preserved series, which at the same time constitutes a gap in characters, can be used as an objective basis for separating species. But the critical problem remains: Continuous lines of descent cannot be divided except artificially. This means that species so divided are not comparable, even by switching dimensions, with those that can be separated at one particular moment of time. They are populations seen historically in the stream of time. Moreover, their status changes by splitting, separation, and divergence as well as by evolution of characters, and no single characterization of a species has more than a temporary validity. The connection of arbitrarily selected "species" in a time sequence, in fact their complete continuity with one another, is to be expected in all evolutionary lineages. But, fortunately, because of the imperfect preservation of fossil faunas and floras, we shall meet relatively few examples of this, no matter how long paleontology continues.

Components of Species. Most species of organisms are composed of discrete individuals, each receiving from one or two parents a quota of genetic information, a genotype, which in turn it may transmit to its offspring in part or virtually as a whole. It therefore carries a sample of the total genetic makeup of the species. The remarks following apply most readily, but perhaps not exclusively, to species that have sexual reproduction and the "gene pool" of which is therefore in a fluid condition. Owing to ecological limitations, species are not universally distributed but are restricted to certain areas, often much more limited than the potentially suitable habitat. Although a species generally is a population of interbreeding individuals, it is not in most cases distributed uniformly throughout its range but for ecological reasons is concentrated in particular localities and scattered more thinly or perhaps absent in other areas between these. Therefore, from the point of view of genetics, local populations are important, for within them genetic exchange occurs more freely than between one such population and another.

A local population has been called a *deme*. This convenient term does not imply that one such population can be distinguished from another in any way except by its locality. Familiar examples are the breeding populations of cold-adapted birds, such as juncos, on separate mountain ranges or peaks, the isolated or semi-isolated breeding colonies of sea birds on offshore islands, and the partially separated populations of desert plants in valleys between mountain ranges of the southwest. Any local population that becomes wholly isolated from others of its species is a potential species itself, for when the exchange of genetic material with other representatives of its kind is impossible, the course of its future evolution can have nothing to do with that of the populations from which it became isolated. As far as the origin of new species out of old populations is concerned, isolation is the critical factor. But a certain amount of divergent evolution can, and often does, take place in a population without complete isolation of its parts.

Within the broad limits just outlined, many species show divisions or components of various kinds that are important for either classification or evolution. In view of much recent controversy concerning these, their meaning should be discussed here. The first category, which is considered by most systematic biologists to have a valid place in classification, is the *subspecies*.

A species may be divided in such a manner that all or most individuals of it in one region are recognizably different from those of another, yet in the zone of contact between these regions there is interbreeding. Commonly the whole range of a species may contain several such distinguishable races. These are subspecies in the geographical sense. It is not yet entirely clear whether it is possible to recognize subspecies that occupy different parts of the available environment in a single geographical area. Such would be called ecological subspecies, but it is questionable whether they could have originated without, at the same time, being geographical.

As far as it serves the convenience of the taxonomist, subspecies receive names. Such a name is a trinomial designating the genus, species, and subspecies together. For example, the white-crowned sparrow, *Zonotrichia leucophrys*, has four subspecies occupying different but adjacent areas during the breeding season. Banks (1964), as a result of examining

2103 "breeding adult specimens" from 62 "populations and groups of populations" gives the following ranges:

> *Zonotrichia leucophrys leucophrys* (Forster).—Eastern Canada east of Hudson and James bays, Cypress Hills of Saskatchewan and Alberta, and mountains of western United States from Montana south to northern New Mexico, southern Utah, central Nevada, and south-central California. Birds recently reported breeding in the San Bernardino Mountains of California are probably of this form.
>
> *Zonotrichia l. gambelii* (Nuttall).—Alaska and western Canada, east to Ilford and York Factory, Manitoba, but intergrading with *leucophrys* south of Hudson Bay from Churchill, Manitoba, to James Bay; south in mountains of Alberta and British Columbia to Hart's Pass, Okanogan County, north-central Washington, and intergrading with *leucophrys* in northwestern Montana.
>
> *Zonotrichia l. pugetensis* Grinnell.—Vancouver Island, from Campbell River, and Vancouver, British Columbia, south along Pacific coast to Cape Mendocino, Humboldt County, California, where intergrading with *nuttalli.* Extends inland into Willamette Valley and to lower slopes of Cascade Range along North Santiam River in Oregon.
>
> *Zonotrichia l. nuttalli* Ridgway.—A narrow Pacific coastal strip of California from Cape Mendocino south to Point Conception, Santa Barbara County, California.

The use of the category *subspecies* is most firmly entrenched in the classification of birds, mammals, and a few other very well studied groups of animals and plants, but it is not obligatory in any case, because the recognition of these divisions of a species depends wholly on circumstances and on the opinion of the taxonomist as to the value of describing and naming them. A subspecies, if it is valid, should be presumed to contain one or more local populations that share certain genetically determined characters by which individuals could be distinguished from those of other subspecies in different areas. If an individual could not be identified to subspecies, except on the basis of locality, then a question could be raised as to the value of naming it. Obviously, this creates one of the major problems in the use of this category, and many taxonomists are satisfied if, in a series of specimens from a given region, 70 percent or more of the individuals can be distinguished from those belonging to a different subspecies in another area. Usually the recognition value of the characteristics of a subspecies is greater than this, but the essential point is not that specimens from a particular locality differ to some extent from those of any other locality but that there exists a range or area within which members of a species all or nearly all differ from others of the same species from some area adjacent to but separate from this. A boundary line could therefore be drawn between the two, not arbitrarily, but indicating the zone in which the two blend into each other. Such blending is commonly in a narrow and rather sharply distinguishable zone, and it is also common to find a natural barrier or at least an ecological transition that coincides with this line: a river, for instance, or a contour level on a mountain range, or the divide between two slopes, such that interbreeding or intermingling among members of populations on either side of the barrier is reduced, although not entirely prevented.

Now, if the circumstances just described apply to some of the populations of some species of animals and plants, there would appear to be an objective reality about subspecies in nature. They can be seen, recognized, and located. They constitute gene pools. But some biologists will not grant the validity of these circumstances, or they object to the concept of subspecies on the basis that it is often very difficult for taxonomists to reach agreement concerning them, that there has been extensive abuse of the naming and describing of subspecies, or that subspecies actually are not entities in the same sense that species are and should not, therefore, have a place in taxonomy. Some of these objections are valid; some are merely matters of differing opinion.

For example, a subspecies is not just a step down from the level of species to something of a smaller scope or distinctness, nor is it necessarily a population that is starting to become a separate species. Rather, the term is applied simply when specimens in a given area have a close similarity to each other and are recognizably different from those in one or more other areas, yet are not genetically separated from those. The amount of difference between members of different subspecies is not important. It varies greatly in different cases. It may apply to a single characteristic or to several, and these may or may not have adaptive value. Not all the characteristics of a species that varies geographically will necessarily vary in a concordant manner or show major changes at the same places in the range of the species, although frequently they do so. This is one of the major arguments of those who object to the use of subspecies in taxonomy. Their case has been well expressed by Wilson and Brown (1953). These authors urge that a thorough study of variation within a species must precede any attempt at taxonomic division or the recognition of races. They suppose that if a subspecies is real it ought to show its differences from other subspecies by concordant differences of any of the characters in which the species varies geographically. Because this seldom happens, they doubt that subspecies are valid biologically or useful taxonomically. This is precisely where the opinion of most systematic biologists differs.

The latter find it advantageous in many cases to base subspecies on one or a few characters, even though they know that the total variability of the species has not been analyzed or included. Their purpose is not the total description of variability throughout the species but the establishment of a convenient term and appropriate description to be useful to those who wish to recognize the known material of a species or, even more simply, to acknowledge in this way the more obvious kinds of variability that a species shows.

On the other side of the picture, there has been a tendency on the part of some taxonomists to carry the naming of subspecies far beyond any possible utility or biological meaning. A recent compilation by Hall and Kelson (1959) shows for one species of pocket gopher (*Thomomys umbrinus*) in western North America no less than 215 named "subspecies." Some of these have been described from a single specimen, many of them from series collected in one or two localities, and others perhaps represent valid geographical races. All, however, add up to an enormous literature which must be consulted by any student who wishes to be well informed concerning the pocket gophers of North America. The question here is whether the labor of analyzing and putting together all 215 diagnostic descriptions of subspecies of this widespread pocket gopher would not be simply a monumental piece of busy work having little use to any biologist. Not all species of pocket gophers are afflicted in this way. *Thomomys talpoides,* for example, has the relatively modest number of 66 subspecies, and so does a widespread species of white-footed mouse, *Peromyscus maniculatus*. Perhaps the time is ripe for a study of variation as it actually occurs in these animals, paying little or no attention to type specimens, diagnoses, and names.

Several other kinds of variation that occur within species have been described and sometimes named as if they were taxonomic categories, although they cannot be so interpreted on any reasonable biological grounds. These will be dealt with briefly in a section following, but one such type of variation, the *cline,* should be mentioned here because it is commonly involved, rightly or wrongly, in the characters of legitimate subspecies. One example out of many is that of the eastern tiger swallowtail butterfly, *Papilio glaucus*. The gradual reduction in size from individuals in the southern states to those in the far north constitutes a cline, that is, a graded change in a single quantitative characteristic with distance. This alone cannot serve to separate subspecies because at no point is the change abrupt. If an east-west line were drawn, say, halfway northward in its range, the individuals south of that line would not be uniformly large, nor would individuals north of that line be uniformly small. It would merely be an artificial and meaningless mark. But in the southern states and as far north as Kansas and Virginia, the females are normally dark blue instead of showing the yellow and black tiger pattern of the males. North of this region, both sexes have the tiger pattern. Whether other characteristics of these butterflies turn out to agree geographically or not, this serves to distinguish two easily recognizable geographic races, *P. g. glaucus* in the south and *P. g. turnus* in the north. Thus a cline may cut across two or more subspecies, and it sometimes happens that several different clines affect different characters at the same time in various directions. Thus it is common in Australian birds for a size cline to run north and south, those northward being smaller, and for a cline in color intensity to run east and west, those to the east being darker. If it should happen that any of these clines are broken by steps, that is, by relatively quick changes in a short distance, these could serve to delimit subspecies, but a continuous smooth gradient of a quantitative character could not be so used, because it shows no clear distinction at any point.

There have not been many attempts to use subspecies in paleontology, for the geographical evidence is seldom adequate to show the distribution of a species as it actually was or the nature of geographical variation at a given time. It would be essential that supposed geographical races occurred at the same time, because otherwise differences that were seen in fossils from different localities could be merely the result of shifting of populations with time, or of local evolution. The attempt to apply the subspecies concept to chronological or successional variations has also been made but seems of doubtful value, for even if the variation itself can be described minutely, we meet the same difficulty as in successional species—separating artificially units that were in fact completely continuous with one another. For those few cases in which this seems to have been done in a useful way, the name "waagenon" was suggested by Caster and Elias. This may avoid some of the faults of such a name as "successional subspecies," but it still designates an artificial unit of doubtful value.

Now we shall discuss certain kinds of variation that do not have taxonomic meaning, although some of them have been treated as if they did. It is unnecessary to consider here in any detail the ordinary intergrading variations of size, height, color, shape, or any other quantitatively varying characteristics that may distinguish individuals but that do not form categories except in a statistical sense. Ordinarily, if such differences are considered by the taxonomist, he has no thought of giving them names or taxonomic rank. Occasionally this may happen as a result of error or lack of information, as when two or more species are named on the basis of specimens that turn out later to be within the range of variation of a single species.

A different kind of variation is that known as *polymorphism,* in which some individuals in a population are distinguishable from others that occur at the same time and place by features that do not intergrade or but rarely do so, and that are genetically determined. Cinnamon bears are of the same species as black bears, *Ursus americanus,* and both may be born of the same parents. Some people have blue eyes, others brown eyes, yet we seldom see intermediates between them and we do not separate them into different species. Tomato worms, which are caterpillars of the tomato sphinx moth, may be either dark brown or green. Females of the common sulphur butterflies, genus *Colias,* are either white or yellow, but nothing in between, and the males are always yellow. In some species of *Colias,* a further polymorphism affects both sexes. The yellow individuals, either male or female, may be orange-yellow or greenish yellow, and intermediates are rare. The white condition of the females has been shown to be due to other genes, more than one pair being involved. Some African swallowtails (*Papilio*) show a polymorphism in which tailless forms mimic species of *Acraea* and differ radically from the nonmimetic tailed individuals. The term "polymorphism" was, twenty or thirty years ago, applied to such cases as these—that could readily be seen and involved morphological features or at least color and pattern. But it now extends to such hidden variants as the blood groups of man and other mammals and birds, in which genetically determined differences occur without intermediates. Perhaps the extreme application of the term is to chromosomal polymorphism, in which recognizably different forms of certain chromosomes occur in fairly constant proportions in the populations of some species, these being due to inversion or translocation of parts of chromosomes. They can scarcely be considered as in the same category as the cases of polymorphism previously mentioned, yet the term has been used by geneticists for a number of years and will probably be continued. A fuller discussion of polymorphism, its genetic basis, and its bearing upon evolution will be given in Chapter 5.

There are some other types of variation to which pseudotaxonomic terms have been applied but which do not come within the definition of polymorphism. One is *seasonal variation.* Many species of insects in warm or temperate climates have several generations in a year. Some of these species show marked differences between successive generations, so that it is possible to recognize, for example, the early spring generation of zebra swallowtail butterflies in the United States by their unusually long tails, small size, and dark color, or to distinguish among various seasonal generations of the small azure butterflies (*Lycaenopsis*) by the amount of black on the edges of the wings. These seasonal forms have been named, and the names are widely used by collectors, but a generation is in no sense a subspecies. The names do not refer to taxonomic categories and should not be mistaken for such. The situation may become even more complicated if such a species is also divided into legitimate geographical subspecies, as is often the case. Here one might be tempted to employ four names instead of three, for the genus, species, subspecies, and seasonal brood. But obviously the latter category does not have taxonomic significance, for the early spring brood is the parent of a late spring brood, and this of an early summer brood, and so on. In such cases it may be legitimate to use names if it is made clear that the names are not intended for taxonomic divisions but for seasonal variations; for example, *Lycaenopsis argiolus pseudargiolus,* "seasonal form *neglecta.*" At this point zoological nomenclature seems to have reached the limit.

A comparable situation occurs in many tropical butterflies, such as the genus *Precis,* that have a generation in the dry season, usually pale in color, and another in the wet season, dark, and sometimes with a quite different pattern. Since the dry-season generation consists of the parents of the wet-season individuals, these two categories cannot be considered races or subspecies or given taxonomic names.

It has been a common practice among insect collectors to describe and name individual variations that appear as single specimens, notable for their rarity and conspicuous because of the difference between them and "normal" individuals. These are in many cases not even genetic mutations, but freaks produced by abnormalities of temperature or mechanical defects in their development or for unknown reasons. Some enthusiasts have accumulated in their collections great quantities of these *aberrations* and have published copiously on them, making the discovery of freaks a seemingly useful scientific occupation. With a little ingenuity, by rearing larvae under controlled conditions of temperature and humidity, it is possible to turn out in quantities such aberrations, especially in the case of butterflies. These are merely individual abnormalities. It serves no valid purpose to name such a creature "aberration *gunderi,*" for example. This can be an entertaining hobby, but science is not well served by creating nuisance names.

To summarize: (1) a *species* is one or more reproducing populations among whose individuals there are either intergrading characteristics or evidence of reproductive continuity; (2) a *subspecies* is a geographically (or ecologically) localized part of a species recognizably different from, yet having reproductive continuity with, other parts of the same species; subspecies are by no means universally present in species of all groups.

MEANING OF CATEGORIES ABOVE SPECIES

When the taxonomist deals with groups of species, that is, with populations among which interbreeding does not take place, his classification uses higher categories, the genus, family, order, class, and phylum, with various qualifying intermediate divisions used when necessary. It is simple, trite, and not quite true to say that a genus is a group of species, a family is a group of genera, and so on. This is often the case, but many a genus is monotypic, as is many a family, and occasionally even an order, because of the limitations of our knowledge. But it is possible to express some criteria for these categories of increasing rank and to demonstrate that their existence is not wholly in the imagination of their makers.

About a dozen species of freshwater sunfishes living in the United States are placed in the genus *Lepomis.* The general characteristics of shape, color, pattern, behavior, life history, and feeding habits are much the same among all these species. The differences between one species and another are greater or more obvious than those between individuals of the same species, and intermediates seldom occur. These species usually, but not always, retain their integrity when living in the same ponds or lakes because their interbreeding is not encouraged by mutual responsiveness, and ordinarily their gametes are not mutually fertile. As a genus, *Lepomis* can be recognized by a combination of characteristics no one of which is necessarily unique to the genus. The anal fin bears three spines, as in some other sunfishes. The scales are relatively large, and this again is true of certain other sunfishes. Teeth do not occur on the tongue, and there is only a small extra bone on the loose margin of the upper jaw, or none at all. A student of fishes can easily recognize at a glance a member of the genus *Lepomis* even if he does not know the species or look at the jaws, the tongue, the size of the scales, or the spines of the anal fin. There is a configuration, a common image, that to him means *Lepomis.*

Such a genus, then, may be characterized in two ways: one, as a bearer of certain structures or characters alone or in combination; and the other, equally valid but less useful, as a group of individuals that differ within certain broad limits of size, form, color, behavior, and so on. That these individuals may be classed as one species or as several within the genus is irrelevant as far as the category of genus is concerned. There may be differences of opinion as to whether these fishes should be placed in one or more than one genus. Taxonomists who have done the latter are regarded as splitters, and they in turn regard the taxonomists of the opposite persuasion as lumpers.

Taking *Lepomis* to be a valid genus, one easily recognizes among other sunfishes (family Centrarchidae) many that are sufficiently different from it to be placed in other genera: *Chaenobryttus, Ambloplites, Acantharchus,* and others. In some cases a characteristic that is unique within the scope of the family may distinguish a genus. For example, *Acantharchus* is the only genus of freshwater sunfish in which the scales are cycloid (that is, without serrated margins). *Acantharchus* is monotypic, its only species being *pomotis,* the mud sunfish living in lowland swamps and streams of the Middle Atlantic states.

The principles here need not be labored further. They are, to a large extent, matters of opinion, custom, and experience of those who study the relationships and classification of animals. It is not necessary, nor is it possible, to justify the values placed upon this or that character as an indicator for the rank of species, genus, family, or higher category. No such justification used in one instance could be applied to other animals or plants that were altogether different in the nature of their characters.

To the systematic biologist of today each category above species means two things: one, a group that is more distinctly recognizable and usually more comprehensive the higher the category to which it belongs; two, it is a branch of its evolutionary tree which, in the case of the higher categories, had a more remote origin, and in the case of the lower categories, a more recent origin.

REFERENCES

Banks, R. C. 1964. Geographic variation in the white-crowned sparrow, *Zonotrichia leucophrys*. *Univ. Calif. Publ. Zool. 70:* 1–23.

Dobzhansky, T. 1951. *Genetics and the Origin of Species,* 3rd ed. Columbia Univ. Press, New York.

Hall, E. R., and K. R. Kelson. 1959. *The Mammals of North America.* Ronald Press, New York.

Imbrie, J. 1957. The species problem with fossil animals, pp. 125–153. In E. Mayr (ed.), *The Species Problem.* Am. Assoc. Adv. Sci., Publ. 50.

Mayr, E. 1942. *Systematics and the Origin of Species.* Columbia Univ. Press, New York.

———. 1948. The bearing of the new systematics on genetical problems. The nature of species. *Adv. Genetics 2:* 205–237.

———. 1963. *Animal Species and Evolution.* Harvard Univ. Press, Cambridge, Mass.

Simpson, G. G. 1961. *Principles of Animal Taxonomy.* Columbia Univ. Press, New York.

Wilson, E. O., and W. L. Brown, Jr. 1953. The subspecies concept and its taxonomic application. *Syst. Zoology 2* (3): 97–111.

3 · REPRODUCTION

Before the processes of evolution are considered in later chapters it may be well to discuss here some relationships between reproduction and evolution, beginning with simple organisms, in which the meanings of reproduction may more readily be understood than in those that are complex. Among one-celled organisms there can be seen a variety of methods of reproduction, usually a self-replication associated with growth. For example, it has been shown that cell division in *Amoeba proteus* is a means of regulating size, for after an interval of growth the animal divides into halves, and growth of the daughter cells continues until each attains the same mass as the parent cell. Further, if a daughter is prevented from reaching this size by removal of parts, it will not divide until the loss has been made up and the mass of the parent cell is reached. Repeated removal of parts prevents division.

Amoeba contains the facilities for carrying on its metabolism and growth as long as it does not exceed a particular mass. The limiting factor here is the ratio of mass to surface area, for as mass increases, surface area does not increase at the same rate. Therefore, in growth a point is soon reached at which the intake and release of substances through the surface could not keep up with metabolic requirements. An *Amoeba* the size of a hen's egg is out of the question. Cell division is then the only way of maintaining life while permitting growth, for the increasing mass is given, by this means, an increment of surface that it could not achieve as a single unit.

As fission, whether binary or multiple, characterizes a great number of microorganisms, it would be surprising if the resultant daughter cells did not sometimes remain in contact, forming clusters, colonies, or multicellular bodies that derive their characteristics partly from being multicellular. Colonial green algae with common sheaths enclosing several cells are examples of this. Likewise, it would be strange if, in some cases, fission did not produce daughter cells that differed structurally or functionally from one another. The gametes (which unite in fertilization) or spores (which do not) seen in various simple plants and animals are examples of such differentiation, sometimes slight, sometimes conspicuous. A method has arisen by which a species survives, not by the potential immortality of all its protoplasm, but by delegating certain parts of this to the function of self-renewal while the remainder eventually dies after having completed the necessary differentiation.

We usually think of the bodies of plants or animals as the primary manifestation of life, the pretext for living, and their evolution is the central theme of the present study. But we might also regard the reproductive cells (actual and prospective) as primary, because it is they that make the unbroken thread continuing forward and backward in time. Thus a body, or soma, is one of an infinite series of houses in which the "germ-plasm" resides in succession. A moment's thought, however, shows that neither the somatic nor the germinal material can be more important, or "primary," but each is equally a means of maintaining a species. Further, it is only in the later stages of embryonic development that any real segregation of reproductive material occurs in most multicellular organisms.

Numerous other devices exist that accomplish the same result as the production of gametes or spores. Individual "bodies" are produced among many multicellular plants and animals by budding or branching (sponges, corals, hydroids, bryozoans, and plants that have vegetative propagation). Multicellular individuals can be produced by fission, as in flatworms, or remain as chains of individuals by fission not quite completed, as in tapeworms. Beginning in algae, the evolution of plants commonly incorporates three reproductive mechanisms: (1) sexual, (2) spore production alternating with sexual, and (3) budding or somatic reproduction associated with one or both of the first two. In all of these, of course, cell multiplication is the major accomplishment, and it is the means of continuing the growth of protoplasm. Each of the principal reproductive methods, except simple cell division, may have originated more than once among primitive organisms.

In Protozoa, it is not difficult to see a relationship among the unicellular flagellates, some of which lack and others which have chlorophyl, or between the latter and the simplest colonial green algae, and thence the Bryophytes (mosses and liverworts). Nor is it difficult to picture an aggregation of collared flagellates by further differentiation leading to sponges. But between Protozoa and such primitive multicellular phyla of animals as the Coelenterata and Platyhelminthes is a gap that few biologists have attempted to bridge. This may be in large part because of a general belief that the matter has already been settled by Haeckel's theory of recapitulation, that because in the development of higher animals a blastula produces a two-layered gastrula, and this in turn develops a third layer of cells, there must have been a similar sequence in the origin of Metazoa.

But there is another possibility, discussed by Jovan Hadži in several papers (for example, Hadži, 1953), and brought to the attention of English-speaking biologists by DeBeer (1954). This is that the turbellarian flatworms are derived from creeping, multinucleate, ciliated Protozoa by a process of partitioning the cytoplasm of the body into many cells at once, without the process of repeated fission that we would otherwise suppose to have happened. Mesozoa may be relicts of the intermediate steps in this episode. No conspicuous enlargement is necessary to bring the bigger ciliate Protozoa within the size range of flatworms, nor need there be any special change in their behavior, habitat, or feeding; given a multicellular body, ciliated and creeping, the digestive canal, proboscis, and simple bilateral symmetry can be regarded as minor modifications of features already well established. The flatworms in turn are considered, in this view, to have given rise to the ctenophores and coelenterates,—the latter by way of the sessile Anthozoa (sea anemones) in which radial symmetry is first established as a specialization. This is in accord with trends seen in other groups of sessile marine animals. (See, however, the full discussion by Dougherty et al., 1963.)

The point of interest here concerns reproductive methods in ciliate Protozoa and in flatworms. Among the ciliates there occur, besides fission, two other processes related to reproduction: endomixis and autogamy. These have been studied in great detail in *Paramecium.* Briefly, endomixis involves a multiplication of micronuclei (which contain chromosomes), a conjugation of two individuals, forming a temporary protoplasmic bridge in which a micronucleus from each crosses to the other and then fuses with a micronucleus in the other, after which the individuals separate and the fused nuclei then undergo division, as the animals do also. This process suggests fertilization, especially as there are in some populations of *Paramecium* several "mating types," and conjugation takes place among these but not between individuals of one type. Autogamy is a process similar to endomixis but takes place in one individual. After their multiplication, two micronuclei fuse with each other in a cone-shaped tubercle close to the oral groove. There is then renewed division of the cell. Neither endomixis nor autogamy is of regular, cyclic occurrence, nor are they essential in the biology of *Paramecium* provided optimum conditions are maintained.

In flatworms (Turbellaria) the body has become more complex, compartmentalized into cells with considerable differentiation of tissues. But as before there are three reproductive methods: fission of the whole animal, sexual reproduction by cross-fertilization (the worms are hermaphroditic), and self-fertilization. These appear comparable to fission, endomixis, and autogamy, respectively, except that (1) more gametes are formed, (2) male and female gametes are differentiated, and (3) propagation is largely delegated to the fertilized eggs instead of employing the whole body. There is nothing here inconsistent with derivation of Platyhelminthes (flatworms) from ciliate Protozoa, and the theory provides a rather simple source for the conditions met in several lower invertebrate phyla, such as Ctenophora, Coelenterata, Mesozoa, Chaetognatha, Kinoderes, Nemertea, and Aschelminthes. It is not, however, wholly established at the present time.

Hermaphroditism may be, then, a primitive condition, a stage of reproduction prior to evolution of separate sexes, and the latter would then have been accomplished in several groups independently: blood flukes (*Schistosoma*) among the Platyhelminthes, coelenterates, and one or more other invertebrate stocks (as well as, quite separately, multicellular plants). As bisexual populations are the general rule in higher organisms, some common adaptive advantage may be found in it. Dispersal comes to mind, yet in blood flukes this hardly seems significant. A mingling of varied genotypes may have long-range adaptive value, providing greater assortment in the gene pool; that is, selection has, so to speak, a choice of the maximum range of variability. There may also be an increased survival rate for a given gene complement, for if a hermaphrodite dies before reproducing, both male and female gametes are lost, but if the death is that of a male, other males remain and the females are not affected, and if of a female, no males are lost. This means that a single death amounts to no more than 50 percent of the gene loss that would result from death of a hermaphrodite, and in practice probably much less.

Further evolution related to sex has produced a multitude of structural, physiological, and behavioral adaptations that ensure the union of gametes and sustain an adequate rate of survival of offspring. The sex of individuals is usually subject to some kind of simple determination between the two alternatives, male and female, by means of the chance distribution of *sex chromosomes*. In fruitflies (*Drosophila*) and mammals the ordinary diploid cells of the body of the male contain one X chromosome and, paired with it, one Y chromosome, whereas in females the diploid cells contain two X chromosomes. At the reduction division when gametes are being formed each gamete of a female will, of course, receive *one* X chromosome, but each gamete of a male will receive either an X chromosome or the Y chromosome, the latter two with equal probability. Therefore, when an egg is fertilized, the two sets of autosomes are brought together from sperm and unfertilized egg, and the X chromosome of the unfertilized egg has added to it either another X from a sperm or a Y from a sperm. Thus the sex chromosomes of the new individual are either XX, producing a female, or XY, producing a male. Apparently in most cases the presence of two X chromosomes is the reason for production of a female, and the presence of only one X chromosome (and not the fact that a Y is also present) is what produces a male. In many cases the action of genes in the X chromosomes is not the only influence resulting in sex determination, but, as in *Drosophila,* it may offset or add to the action of genes carried in the autosomes. In fishes, birds, and moths an opposite system is used, in which the female gametes are of two types but the male gametes are all alike.

Certain other animals have an altogether different method of sex determination. The female of *Bonellia,* a marine echiurid worm, lives in mud under the sea, with its proboscis extended up into the water. The male is a microscopic ciliated creature living within the female's body. The larvae, like those of most marine worms, are ciliated and free-swimming. Sex is undetermined during the swimming stage (or, more accurately, the larva is a prospective female but subject to reversal). If the larva settles down in the mud it becomes a female, but if instead it first comes in contact with the proboscis of a female worm, it then enters and becomes a male inside the reproductive tract of the female. Evidently a chemical influence of the female upon the larva suppresses the latter's female potentiality, which would otherwise determine its course of development.

Crepidula (the slipper shell, a marine gastropod) develops, after a free-swimming larval stage, on a hard surface, usually a rock or another shell. Sex is again potentially female, but if a larva settles within a few centimeters of a female, its potential femaleness is inhibited by a secretion from the latter and it becomes a male. Beyond a short distance the concentration of the substance in the water is too low to produce this effect. The adaptive value of these devices in *Bonellia* and *Crepidula* is, of course, to increase the probability that the eggs of the female that is already in place will be fertilized.

Another modification of the common male-female determination is that in which some or all of the eggs start development without fertilization. In honey bees (*Apis*) nonfertilization results in males, the drones of the hive, while fertilized eggs produce females that are either reproductive (queens) or sterile (workers), depending on the nutrition they receive while larvae. In many species of aphids in temperate countries males occur only in a generation at the end of the summer. The fertilized eggs from this generation pass the winter, hatching next spring into females only, and these reproduce parthenogenetically, several generations of females following one another through the summer. Among some tropical aphids there may be no males at any time.

The list of remarkable adaptations for reproduction is long, as special circumstances make for selective pressure in unexpected ways. But two primary functions are served by whatever process of reproduction we may consider. One is to transmit genetic "information" to the organisms of the coming generation, and the other is to start a succession of developmental processes under control of this information that will produce the structures and functions of a new individual. The last 60 years of research in genetics have provided an enormous fund of knowledge of the basis of heredity, and this is equally true of the complex problems of embryonic development. Recent investigations are beginning to show how the genetic material of the fertilized egg controls differentiation in the zygote, the embryo, and later stages of development.

REFERENCES

DeBeer, G. R. 1954. The evolution of Metazoa. In J. Huxley, A. C. Hardy, and E. B. Ford (eds.), *Evolution as a Process,* 2nd ed. G. Allen & Unwin, London.

Dougherty, E. C., et al. 1963. *The Lower Metazoa. Comparative Biology and Phylogeny.* Univ. California Press, Berkeley. Dougherty, E. C., Z. N. Brown, E. D. Hanson, and W. D. Hartman (editors).

Hadži, J. 1953. An attempt to reconstruct the system of animal classification. *Syst. Zoology* 2(4): 145–154.

4 · HEREDITY

HEREDITY AND ENVIRONMENT

If all individuals of a species were precisely alike there would be no study of heredity, no problems of genetics, for each generation would be a copy of the preceding. But even "identical" twins are not copies; more important, any particular feature of one individual differs from the corresponding feature of another. Some of the variations in these are transmitted to offspring in apparent disregard of circumstances; white forelock (a white patch in the hair over the forehead) appears in some of the children of those who have it. Short fingers (brachydactyly) resulting from fusion of normally separate bones are inherited in a similar way. The fact that we observe the inheritance of such clear-cut examples of *unusual* variant characters leads us to infer that the mechanism that controls transmission of these must also control that of the *usual* characters (hair without a white patch, fingers of normal length, and so on) to which we would otherwise pay no attention.

Besides these traits that seem to be entirely hereditary there are others, such as a well-tanned skin, inch-long fingernails, or the ability to speak French, that are induced by the action (or absence) of some influence in the environment. Yet there are many cases in which it is not easy to distinguish the inherited traits from those that are acquired. If corn plants, known to have uniform heredity, are grown under nearly identical conditions they reach approximately the same height at the same time, but if some of them are subject to dissimilar soil, temperature, or water supply their growth is varied accordingly. What the organism inherits, then, is an ability to grow at a certain rate, to produce certain characteristics, under a given set of conditions.

Among the obvious characteristics that an organism shows, there may in some cases be two or more alternatives that appear in response to normal but contrasting environmental conditions. The plant called mermaid weed (Fig. 6) grows so that its leaves are partially in and partially out of water. Those that grow below the surface are finely divided, and yield easily to the current, so that the plant is not dragged or pushed. Those that stand above water level are complete and have a larger surface area. Obviously both types of leaves are adapted for maximum usefulness in their situations, and both carry on photosynthesis. The plant might be said to take advantage of its two worlds, above and below water. Its heredity permits these two quite different responses during growth and differentiation of the leaves.

Both inherited and environmental factors enter into the development of any trait, although for practical purposes we usually treat characteristics of an organism as if they were controlled by one or the other. A man does not develop a well-tanned skin if his skin is unable to respond to ultraviolet light by producing pigment; he cannot speak French if heredity has failed to provide him with the physical equipment for speaking at all or the mind to learn a language. At the other extreme, the simple hereditary control of eye color in fruit flies hinges upon the provision of fruit pulp, water, air, and a fairly narrow range of temperature for development of the larvae to and beyond the pupa stage, in which the adult eyes will be formed. These considerations warn against thinking of heredity and environment as contrary forces.

MECHANISM OF HEREDITY

It was not until the last two decades of the nineteenth century that tangible evidence of the mechanism of heredity began to emerge. Van Beneden, Hertwig, and others found that the nuclei of plant and animal cells contain bodies that can be seen when stained dark, the chromosomes. These retain their form and number with remarkable stability and are alike in cells of corresponding parts in all individuals of a species, although more easily distinguished during cell division than at other times. Inasmuch as a fertilized egg divides and redivides until thousands or millions of cells result, each chromosome must do this also at each cell division. All the components of every chromosome must also duplicate themselves with great precision and continue to do so rapidly as long as the cells divide.

In virtually all animals and plants, whether hermaphroditic, parthenogenetic, or otherwise, at least one full set of hereditary factors must be transmitted to the offspring by each parent. It soon became clear that the behavior of chromosomes in the gametes (a single complete set), in the fertilized egg (two matching sets), and in the cells that develop from the egg (the same two sets, formed by replication every time a cell divides) corresponds exactly with that which would be expected if they carried the genes. In order for fertilization to bring together two full sets and no more, it is necessary to interpose, in the production of eggs and sperms, a cell division called meiosis, in which the two sets (diploid number) are merely separated from each other, giving the gametes the haploid or monoploid number, one set each. Many of the simple facts of genetics are explained and are predictable from this manner of transmission of chromosomes.

Numerous kinds of organisms, especially among plants, are not so limited in their chromosomal arrangements but have triploid, tetraploid, or higher numbers of sets. Of several species in a genus of plants it often happens that the distinctive characters of some species are traceable to the polyploid condition; for example, various species of wheat, cotton, grasses, and so on, have different multiples of the haploid number of chromosomes. Comparable in a way to this polyploidy is the structure of the "giant chromosomes" in the salivary gland cells of *Drosophila* larvae. These chromosomes, many times larger than in other cells, consist of numerous parallel threads, each evidently representing a single chromosomal unit (chromonema). It is normal for any chromosome to replicate itself, but in this case many replications occurred without division of the cells.

Genes and Alleles. For simplicity we may imagine one chromosome as a microscopic thread consisting of a sequence of genes, each one of which we could represent by a letter: ABCDEFGHI. . . . Each of these is an independent "starter" of processes that result in producing one or more of the hereditary characteristics of the organism. As chromosomes usually occur in homologous pairs, there should be another to compare with the one already imagined, and on it there should be the same genes, in the same sequence. But this is not quite true; genes at a given place on a chromosome (locus) affect the same characteristic but not necessarily in the same way. For instance, in snapdragon flowers a gene at the A locus on both chromosomes of the appropriate pair could result in red flowers, and as we know of its existence only by the visible result (phenotype), we call it the gene for red. But perhaps the gene at the same locus on the second chromosome is not for red, but when present along with *A* the two genes together cause the flowers to be pink; this combination (genotype) can be represented simply as *Aa*. What would happen if the genotype were just *aa*? The flowers in this case are white. By means of these three genotypes, *AA*, *Aa*, and *aa*, three different colors appear, although only two contrasting genes are involved. Gene *A* and gene *a* are *alleles*.

For some genes three, four, or even larger numbers of alleles are known, each having different effects. But no matter how many, it is not possible for more than two to be present at a time in one diploid organism because, of course, the chromosomes are in pairs. Every pair of chromosomes (4 pairs in *Drosophila*, 23 pairs in man) carries its own different sets of genes. Each locus on each pair has its own particular gene. For many of these, alleles are known, and they occur with varying frequencies in a population but no more than two for any diploid individual.

Dominance. In the example just given of the color of flowers in snapdragons only two alleles were present, *A* and *a*. The homozygous *AA* produces red flowers; the homozygous *aa* produces white, and when the two are combined in a heterozygous genotype, the color is intermediate, pink. Evidently then, the *A* and *a* contribute about equally to producing the intermediate condition. Neither one can be said to be dominant over the other. This is often true of alleles, that each contributes its share to the resulting phenotype, but it is even more common for one allele to suppress, partially or entirely, the effect of the other. So, for example, in Mendel's experiments with peas, he found that in a cross between homozygous smooth peas and wrinkled peas, all the offspring (heterozygous) were smooth. Representing these genotypes as $SS \times ss \rightarrow Ss$, we see that even though the heterozygote is smooth in appearance and cannot be distinguished from the homozygote, *SS*, nevertheless it has a gene for wrinkled that is not expressed. We therefore refer to *S* as the dominant allele and *s* as the recessive. This is all that these terms mean in genetics: the ability of one to conceal the effect of the other, or to prevent the other from having any effect. There are degrees of dominance; it may be complete or partial, but it does *not* refer to personality, strength, or adaptive value, as a nonbiologist might think. Also, there is nothing to cause a dominant gene to spread through a population more rapidly than the recessive does. As long as mating is random, and one allele has no greater selective value than the other, they do not change in frequency relative to each other.

Mutations. In spite of the remarkable constancy of genes through generation after generation and the astonishing accuracy of their replication with each cell division, occasionally there is a change in what a gene can do. This is known only by the fact that it has made a difference in the phenotype, for we cannot look (yet) at a gene. This change, occurring at one locus occupied by one gene, is a mutation. It alters the chemical nature of one gene on that chromosome, and thereafter this gene is replicated in the daughter chromosomes of daughter cells in the altered form. The mutated gene now is an allele of the older, nonmutated gene already present at this locus in other cells.

It is not yet, but soon may be, clear what chemical change takes place at the locus of a gene when it mutates. Our present use of the word *mutation* therefore must refer to the sudden appearance of a new phenotypic character (or more than one if the gene has multiple effects) in an individual among whose ancestors it did not occur. There are, however, some unusual phenotypes that we know did not originate from a mutation at a single locus. De Vries, a Dutch botanist of the late nineteenth century, used "mutation" for a new variety of evening primrose that appeared in his greenhouse and was so distinct from others as to be described as a new species, *Oenothera lamarckiana.* It has since been found that this was tetraploid (having four instead of two sets of chromosomes), and the effects of having different proportions of alleles from those in a normal diploid plant were responsible for the strange new phenotype. Such changes may also be brought about by a variety of chromosomal abnormalities, such as translocation of parts, duplication or deletion of parts, or various polyploid conditions. Since these cannot be determined except by cytological study, it may be necessary to use "mutation" to include the results of these aberrations as long as this study has not been made. But the word now more generally means a change in the effects of a particular gene.

Almost any kind of conceivable phenotypic change may occur as a mutation, subject only to the limitations of the material involved in producing the phenotype. For instance, one of many genes affecting the wings in *Drosophila* causes a short stump to form instead of a wing, and this, of course, is useless for flying; if it took place in nature instead of a laboratory cage the insect would probably be picked up by a predator. But under artificial conditions we can maintain it and determine the method of inheritance. Another mutation causes normal wings to appear on two segments of the thorax instead of on one, so that a four-winged fly results. Numerous mutations of the color and form of the eyes have taken place in *Drosophila,* such that a long series of color alleles exists having various degrees of dominance or recessiveness to one another and to the original wild type. By means of other genes eyes may be reduced, bar-shaped, or absent, because of past mutations. Any of these and many other known mutations may of course be perpetuated by breeding in the laboratory, provided it does not also interfere with the development or survival of the fly.

For some mutations the phenotypic effect may be attributed to a change in the rate of action of a process that is involved in making the phenotype. For instance, a color mutation in goldfish depresses the rate at which pigment is accumulated in color cells of the skin, so that during the normal lifetime of the fish the new allele in homozygous condition prevents development of more than a small amount of scattered pigment, leaving the fish nearly white. When this genotype is heterozygous, the color is intermediate between that and the normal gold. Undoubtedly a great number of genes have such effects on the rate of a process, to increase or decrease it, and this would be especially noticeable in growth of the body or of its parts.

Commonly mutations may be lethal, causing death in some stage of development between fertilization and the adult, or are semilethal, resulting in premature death or lowered viability of an individual. Hemophilia, a congenital disease in which the blood does not clot easily, is an example of such a semilethal gene in man, for it makes survival to maturity extremely unlikely; one may bleed to death from a small cut. The gene is recessive and its effect is seen in males primarily, because it is carried on the X chromosome, and in males there is only one of these, so the normal dominant gene cannot be present to prevent the effect. Most lethals and indeed most mutations of any kind are recessive to the "normal" gene from which they arose. Consequently, they may be concealed or only slightly expressed in the heterozygous condition, and therefore can be spread by random breeding in a population for many generations even if they are lethal or otherwise maladaptive, because their effect will not appear except when an individual receives two of the recessives (homozygous) because each of its parents happened to be heterozygous. Even then the result may not be apparent, because if the gene is lethal the individual dies, perhaps before birth or hatching, for no obvious reason, and is not noted.

More often than not, mutations are at some selective disadvantage, either in survival or in reproductive capacity. This might be expected when we think of the multitude of adaptive adjustments made during the long history of any species, by which selection has established thousands of mutually coordinated phenotypic effects in any average, normal individual. The probability that any change at all in the total phenotype would be an improvement over the normal, or in other words that a random change in the effect of any gene could increase its selective value, is very small indeed compared with the opposite. But some mutations are neutral, and although they make a perceptible difference in the phenotype, this has neither a greater nor a less selective value.

Many hundreds of mutations are known to have occurred during the course of genetic studies of *Drosophila* (in which millions of individuals have been reared and observed) and many more in domestic animals, cultivated plants, and man. The effects and methods of inheritance of these are in general well known. It is clear that the mutant gene, if it is carried along in the population at all, constitutes an allele of that which gave rise to it. Conversely, a great many alleles, known from breeding experiments, must have originated at more or less remote times from other genes by mutations. As we look back over the early history of any lineage of organisms and see the slow accumulation of new characteristics in structure, development, physiology, and behavior, along with the loss or replacement of old traits possessed by their ancestors, it becomes obvious that the action of any gene now existing must be the result of a succession of mutations of antecedent genes, occurring perhaps at long intervals and supplanting their predecessors because of some selective advantage. This subject will be discussed in later chapters from the point of view of populations and selection.

The kind and degree of phenotypic change by mutation should be mentioned. Usually mutations are slight, such as a minor increase or decrease in rate of growth or of a physiological process or an act of behavior. Teeth may grow for slightly longer time or slightly faster than usual. Cusps on molar teeth, or perhaps just one cusp, may come to differ in a measurable, although small, degree as a result of mutation. Sometimes mutations are much more obvious than this, as the failure of the last molar in upper and lower jaw to appear at all, even as a bud or, again, the development of additional supernumerary teeth for which there is no available space in the series. Occasionally mutation produces a grossly abnormal or defective individual, presumably by interfering with the course of development at an early age. This does not mean that all gross abnormalities are caused by mutation, however, for those produced by nonhereditary causes are at least as common. Physiological or chemical mutations are frequent; the allele responsible for sickle-cell anemia, a blood disease in some people of tropical Africa, undoubtedly began as a mutation.

We should not, however, expect too much to be accomplished in this way. A mutant plant may have very different leaves from others of the same species, but they are still leaves, or structures comparable with leaves, perhaps with certain tissues modified, or chlorophyl lacking, or a new pigment introduced, or a hairy surface, or a thickened or thinner cuticle, or a shape of a tendril or a thorn. But a mutant gene in a plant cannot cause it to grow wings instead of leaves, or make a bud develop into an arm and hand or a root become a caterpillar. Such changes may not be beyond our imagination but they are certainly beyond the capacity of a growing plant, which must use the materials obtainable by it and within its chemical and mechanical limitations. A gene is a part of a constellation of interacting genes. The total genotype is already highly adapted, by way of the integrated phenotype, to special ecological situations; no gene is free to produce its own effects independently of relationships with the rest of the organism.

Some biologists have suggested that the origin of major groups of animals and plants, such as phyla, classes, or orders, may at times have come about through single mutations involving large and complex changes that happen to be successful. Such creatures, called by Richard Goldschmidt "hopeful monsters," seem, with the advance of our knowledge, to be less and less necessary to explain the beginnings of new adaptive advances. A monster is far more likely to be hopeless than hopeful. The gaps formerly present in our knowledge of many groups are being filled today by evidence of the usual, gradual transformation of characters under the action of natural selection.

Sources of Genetic Variability. Only those kinds of variation that originate in genetic changes and are transmitted by some genetic mechanism can be said to evolve. Variability caused by environmental action on individuals is not known to be inherited and does not affect the course of evolution. But there are several categories of genetic variation. One is the mutation of single genes, a rare and entirely unpredictable event. Neither the occurrence, the causes, nor the effects of mutation in nature can be anticipated. We know in retrospect that the mutation of certain genes (as for instance the change of a dominant gene for color to a recessive for albinism) may take place in a wild population more than once, perhaps many times in the course of years, giving the same result. But these events are still rare, far apart, and unpredictable. Also it is possible for a reverse mutation to occur in an allele that itself originated as a mutation.

It has been shown that a number of different agents can increase the frequency of mutations: radioactivity (ionizing radiations—alpha, beta, and gamma rays—X rays, protons, and neutrons), the rate produced being in proportion to the dose; various compounds (phenol, formaldehyde, urethane, peroxides, and so on, including certain substances that induce cancer). The mutation rate can be most simply expressed as the number of mutations at a particular locus per gamete, and this differs for various loci as well as different kinds of organisms. The frequency of new cases of chondrodystrophic dwarfism (a dominant mutation, new cases being those not involving parents or known ancestors) in Denmark is about 1 per 12,000 individuals; inasmuch as either of two genes at a given locus (chromosomes being in pairs) may mutate, the ratio is actually about 1:24,000. This is still only an estimate, for we cannot be sure that only a single locus is capable of producing this mutation. Estimates for other mutation rates vary from 1:10,000 down to 1:100,000 or less. Surprisingly, there is evidence to indicate that some genes are mutagenic. That is, when present they are associated with a higher rate of mutation of other genes.

It has become clear that most mutations (that is, most genes) have more than a single effect. A mutation produces, besides a modification of the gene effect already known, some other abnormality or variation that seems to have no reasonable relationship to the phenotype already known. Scott (1943) found that mutations of the majority of supposed single-effect genes in *Drosophila* caused, in addition to structural differences, perceptible changes in behavior. This, of course, gives a clue to that part of behavior which was controlled by the gene before mutation. Multiple effects are probably characteristic of most genes. This is *pleiotropism.*

A mutation in a cell within the body cannot be expected to affect that individual appreciably, but if it occurs in a prospective or actual gamete, and this is transmitted in fertilization, then the mutant gene will appear in all cells of the new individual and may show its phenotype (unless concealed by recessiveness). There are, indeed, somatic mutations that occur in a part of an organism that is developing, and among plants these sometimes affect the appearance or function of the parts that ultimately develop from the cell where the mutation occurred. If it interferes in some way with the function of reproductive organs (flowers, fruit, or seeds) developing on that branch of the plant, then it may still be possible for the mutant branch to be grafted or propagated by cuttings. This was the case with the seedless navel orange. It is difficult to see how somatic mutations could have any effect in nature on plant evolution unless the plant is one that reproduces vegetatively, as some kinds do.

Other sources of genetic variability besides mutation depend in part on the fact that alleles are present as a result of past mutations. This means, for instance, that in fertilization of an egg homologous sets of chromosomes brought together do not merely duplicate each other, but introduce into the new zygote a certain proportion of contrasting alleles derived from the two parents. But because of dominance effects, the result will not necessarily be a close replication of the phenotype of either parent. The genes, then, are reassorted with each new act of fertilization, and the phenotypes differ in ways that cannot be anticipated except in reference to particular genes that were known to be present in the breeding pair.

In organisms with sexual reproduction there is yet another source of genetic variability in the "crossing over" that occurs between homologous chromosomes just prior to their separation in the meiotic or reduction division before the gametes are formed. This is an exchange of parts that occurs in a random manner in a pair of chromosomes. They become attached to each other throughout their length, each section of one with the corresponding section of the other, and then separate, but if there is a crossover, this means that at some point the lower portion of one has remained connected with the upper portion of the other and vice versa. Therefore, a series of genes that happened to be present in the lower part of one chromosome is henceforth carried as a part of the other chromosome instead; thus when a gamete receives a chromosome of this pair the alleles in it are, in part, not the same as they would have been if there were no crossing over. Crossovers may be double, or even triple, and in each case unpredictably.

The consequences of this *recombination* are, when added to the reassortment in fertilization, enough to ensure a virtually infinite number of changes in the precise pattern of alleles received by any population in the course of one breeding season or generation. Since, except for mutation, these variations are related to sexual reproduction, and since they provide a constantly changing spectrum of phenotypic characters in new individuals as these appear, there is an immense store of variation from which natural selection may save the more viable and more adequately reproductive phenotypes of coming generations. We might almost think, as several biologists have suggested, that sex is an adaptation, brought about by natural selection, to provide and promote variability in order to aid in the adaptive success of populations. This raises a question, however, as to any way in which natural selection (acting only on what exists at any particular time) could develop mechanisms that produce variability in anticipation of some future advantage to be gained by it. Is it not more likely that genetic variation is an incidental advantage, as we see it in retrospect, rather than the primary value that led to evolution of sex in the first place?

THE GENETIC MATERIAL

Within the nucleus of a cell the primary genetic material that comprises the chromosomes is deoxyribonucleic acid (DNA). Each molecule of this is an incredibly complex chain of, usually, many thousands of nucleotide units. Another compound, ribonucleic acid (RNA), is present both in the nucleus and in parts of the cytoplasm, especially the ribosomes. Various proteins may be associated with these nucleic acids in combinations called nucleoproteins. Any protein molecule is made of a great number of amino acid units, each one of which, by itself, would be a relatively simple compound. At least twenty different amino acids are present in the proteins of cells, but in varied combinations.

The nucleic acid component is also made, as noted above, of one or more chains of nucleotides. Each nucleotide unit contains a sugar (either deoxyribose, in DNA, or ribose, in RNA, but not both), a phosphate group (PO_4H^+, or with calcium, sodium, or magnesium in the place of hydrogen), and a nitrogen group in which are either one (pyrimidine) or two (purine) carbon–nitrogen rings (referred to as "bases"). The pyrimidines in DNA are cytosine and thymine, but in RNA they are cytosine and uracil. The purines in both DNA and RNA are adenine and guanine.

Thus the DNA strand is made of thousands of nucleotides in a chain, each of which has in it cytosine or thymine or adenine or guanine (bases). That is, for the position of any nucleotide unit, one base of the four is possible. For two adjacent nucleotide positions in the sequence there are two bases present, each of which is of four possible kinds; this means there are 16 (4^2) possible combinations of any two adjacent bases. In any three adjacent nucleotide positions the number of possible base combinations is 64 (4^3). Likewise, if n is the number of nucleotides in a chain (say many thousands), then 4^n is the number of possible combinations of these bases, taking the chain as a whole.

But in the DNA molecule there are two parallel strands or chains, both spiraling in the same direction and connected with each other as the two sides of a spiral stairway are connected, by steps or rungs. Each rung consists of two complementary bases; that is, a base in a nucleotide in one strand is connected with a complementary base opposite it on the other strand. One of the bases is a purine, and the other (in the same "rung" of the ladder) is a pyrimidine. Adenine is the complement of thymine and guanine the complement of cytosine. One complete twist (360°) of the spiral ladder or helix has ten of these base pairs, 3.4 angstroms apart, and is thus 34 angstroms long, each rung turned at 36° from the next. At least two H bonds link the two bases of each pair or rung, and the width of the ladder, from side to side, is 11 angstroms. An angstrom is one hundred-millionth (10^{-8}) of a centimeter.

One strand of the double helix can only be separated from the other by unwinding the spiral and breaking the bonds between the complementary bases, but this is done at each replication of DNA. A new strand can be constructed by each of the separated strands because the broken rungs attach to themselves nucleotides containing, in each case, bases complementary to those already present. These are then connected with one another in series, making a new strand, with the help of enzymes. In a similar way, it is thought, DNA can construct RNA strands that are like mirror images of itself, and the RNA, separating, can act as a source for construction of proteins either in the nucleus or, by transfer, in the ribosomes. This is the means by which clues or messages are given for initiation of particular sequences of amino acids in building new proteins. Since the combination of bases in sequence along one strand serves as a template (like a negative) for the combinations of amino acids that are complementary to them or that are, so to speak, selected by them from the surrounding cytoplasm, these base sequences provide the "genetic code" for starting development of new proteins of specific kinds, including enzymes, which are themselves proteins. A long series of nucleotide bases in threes, for example, is sufficient to convey the message leading to construction of a particular protein out of specified amino acids. As four different bases are available, and as any three of them in any combination are able to specify an amino acid, it is evident that the code message is comparable to the varied spelling of three-letter words using an alphabet of four letters. The number of possible "words" is more than enough to designate twenty different amino acids.

Proteins antecedent to hemoglobin, for instance, or hemoglobin itself (which has several slightly different variants) can be constructed by way of a message conveyed by RNA to the ribosomes where this is done, and a single variation in a base at one point of the series can be enough to change, in one section of the hemoglobin molecule, a component amino acid. This is the nature of biochemical mutations, as is now seen dimly in the rapid advance of our knowledge of the genetic code.

Although recent authors have rightly described these new discoveries as among the greatest scientific advances ever made, it is well to add that at our present state of knowledge they have only an indirect bearing on evolution. Our knowledge of the material itself does not yet affect appreciably our understanding of the evolutionary process, which is a slow change in successive populations of organisms under the influence of natural selection.

REFERENCES

Scott, J. P. 1943. Effects of single genes on the behavior of *Drosophila. Am. Naturalist 77:* 184–190.

Strickberger, M. W. 1968. *Genetics.* Macmillan, New York.

5 · POPULATIONS

The concept of *population* as distinguished from species or subspecies has come to take a central place in the thinking of evolutionary biologists. But, apart from differences of emphasis or interpretation, the concept is plastic, because relationships among individual organisms are so diverse that they cannot be simply defined. The value of the idea of population is that it expresses the most direct hereditary and reproductive relationships among individuals and that it distinguishes a kind of evolutionary unit. It is the visible manifestation of the "gene pool." By the broadest definition, a population is a group of individuals of one species so situated that interbreeding can take place among them and which thus has continuity through successive generations.

Although most kinds of organisms have such populations, many do not. There is no interbreeding or genetic exchange between individuals of asexual species of animals or plants. The only sense in which the term "population" might apply to them is that of hereditary continuity with offspring and with ancestors, for their reproduction is by individuals acting alone. In many kinds of aphids a generation with sexual reproduction is followed by several or many generations that have no genetic exchange, being parthenogenetic females. Among the social termites, ants, bees, and wasps sexual reproduction is limited to a small minority of individuals, and the rest are sterile. It seems reasonable, then, to extend the idea of population to include all individuals that occur in colonies whose reproductive members are able to interbreed. But the word "population" in the sense of genetic exchange has no meaning in species that reproduce only asexually, inasmuch as individuals then have nothing to do with one another reproductively except as parents or offspring. The study of genetics in such cases would be concerned with clones, as already indicated, and the material of variation in these is produced by mutations, but not by recombinations of genotypes.

Among biparental organisms, the characters of the offspring are influenced by two or more sets of genes and by the complex interactions that occur among these genes. In each generation the genes undergo reassortment during the process of gametogenesis and fertilization, and therefore each set of chromosomes in an individual, although homologous with a similar set in the parents, has its own distinctive combinations of alleles. These are delivered to and again reassorted and recombined in its offspring. Thus, a population because of the interbreeding within it constitutes a reservoir or "pool" of genetic material that is continually in the process of rearrangement. Most of this rearrangement is sufficiently random and uniform to result in no significant change in the proportions of various genes in the population or in the characters of the organisms. But over a period of time, largely because of the influence of selection, such changes do occur, and this is evolution.

It is necessary now to consider the kinds of populations and their relationships to each other. Obviously any single locality or any habitat may contain from a few to many thousands of populations, most of which belong to different species of organisms. Our discussion need not concern relationships among all these but only among the populations that lie within the limits of possible interbreeding with one another (that is, a species). There are several characteristics of populations, each highly variable independently of the others.

GEOGRAPHY

One such characteristic is the geographical relationships that a population has with others. Occasionally there is only a single population of a species, within which interbreeding can be carried on and which is not divided into smaller, local demes. It is actually a single deme itself. This must usually be the result of an enclosure of the species by an ecological barrier that individuals cannot pass, as on an island surrounded by water or a pool surrounded by dry land or a cold mountain top surrounded by warm lowlands. For example, the entire population of the White Mountain butterfly, *Oeneis semidea,* occurs above timberline on the Presidential Range in the White Mountains in New Hampshire. The restriction was caused by a warming of the lower land following the retreat of the glaciers, so that the cold-adapted insects are limited more and more to the higher peaks, although they were formerly widely distributed across the northern parts of the continent, as is shown by several other species of *Oeneis* that occur in Canada and in mountain ranges of the west.

There is a species of desert killifishes, *Cyprinodon diabolis,* "confined to a single cave-spring hole in Ash Meadows, Nevada. . . The total population of this species probably varies between 100 and 300 individuals. . . This species has been able to maintain itself over many thousands of years because of the isolation of its habitat in a remote desert area" (Miller, 1961). The restriction originally must have resulted from climatic change, causing a discontinuity in streams. In the same paper many other instances are given by Miller of fish species in the southwest being isolated or greatly reduced in range, either as originally discovered or by environmental disturbance caused by man in the last century. Several such populations have been destroyed and a few species wholly exterminated. A frequent cause is the well-meant introduction of exotic species, such as the carp, which has heavily damaged various native fishes in the west.

Many other species of animals and plants are in a stage of diminishing ranges and consequent discontinuity of their populations, either as a result of slow changes of environment or because of the action of man or other competing species in reducing the available area in which they can live. *Desmognathus wrighti* is a small brown salamander living under stones and logs in the higher parts of the southern Appalachian Mountains. There is a colony on Mt. Mitchell, another on Grandfather Mountain, and other isolated populations on Mt. Leconte and a few other peaks; but there is no connection between these populations. As yet, however, there is no report of differences in their characters.

Even more common are examples of widely distributed species in which the largest number of individuals occur in extensive populations that may be moderately localized or scattered with occasional communication between them by wandering individuals and, in addition, numerous outlying disjunct populations here and there in favorable locations which are beyond possible communication with the rest. This appears to be the case with the mountain gorilla in the eastern part of the Congo and Uganda, as shown in a study by Emlen and Schaller (1960). It is also true, although less completely mapped, in several species of salamanders in the eastern United States, such as *Plethodon cinereus* and *P. glutinosus*. Of course, any population that becomes disjunct (isolated from others) has lost the genetic control that would have been exerted on it by interbreeding and therefore inevitably accumulates its own characteristic spectrum of genotypes. To some extent this is true also of populations that are not entirely disjunct and between which a small amount of gene flow takes place. It is seldom possible to predict what will happen in the future dispersal and contacts of these populations, but probably the presence of a number of disjunct "island populations" around the principal range of a species is an indication of reduction from a formerly greater distribution, as is unquestionably true of the gorillas in Africa.

Among a great many kinds of animals and plants a species may occupy a very wide range with what seems to be fairly uniform density of population. For example, throughout the well-watered parts of the eastern United States, meadow frogs, *Rana pipiens*, can be expected in the vicinity of pools or streams or along the edges of lakes. These frogs are active during several months of the year and are quite mobile, so they disperse fairly readily and the population of a small pond is not likely to remain outside the genetic influence of other populations living along streams that run into or out of the pond. It is true that the breeding congregation in early summer is somewhat more localized than the distribution of the species at other times, but a frog hatched along one part of the shore of a lake does not necessarily return to the same spot for its own breeding.

In a great many kinds of animals the breeding range is only a small fraction of the area occupied by the species throughout the year. For example, fur seals range widely over the Pacific Ocean but return for their breeding season to a few limited localities on islands off the coast of Alaska, where they climb out on shore and establish large colonies. It is here and only here that breeding populations can be distinguished. Much the same is true of various Pacific Coast species of salmon that come upstream in the spring to spawn in tributaries where they were originally hatched, in the meantime having gone down to the ocean and there mingled with other fishes of the same species that come from other rivers and tributaries. A breeding population is then fairly well distinguished by the particular part of a river system to which it returns, for it is only within this area that interbreeding normally takes place, and such an area may be separated from others by long stretches of river that are unsuitable for spawning. The word population in the evolutionary sense, as already discussed, therefore refers in these species only to the population present at the time and place of breeding.

Another example of the same principle is seen in migratory birds that return year after year to particular places for nesting. Regardless of the winter ranges of song sparrows, for example, their biological populations are those that can be distinguished in the nesting season. The importance of this is that because isolated breeding populations may undergo independent evolution, they have a potentiality of becoming species. Various populations, subspecies, or species of sparrows may, although closely related to one another, have different breeding ranges and yet occupy the same or overlapping winter territories (Fig. 1).

GEOGRAPHIC VARIATION

Recent studies of the development, fertility, behavior, and other characteristics of *Rana pipiens* in various parts of its range (Moore, 1949, 1950; Ruibal, 1955) have shown numerous differences in all these features. When frogs from the southern and northern parts of the range are brought together for breeding, the differences are great enough to affect fertility and to cause complete sterility between the more northern and southern members of the same species. Thus even without the exist-

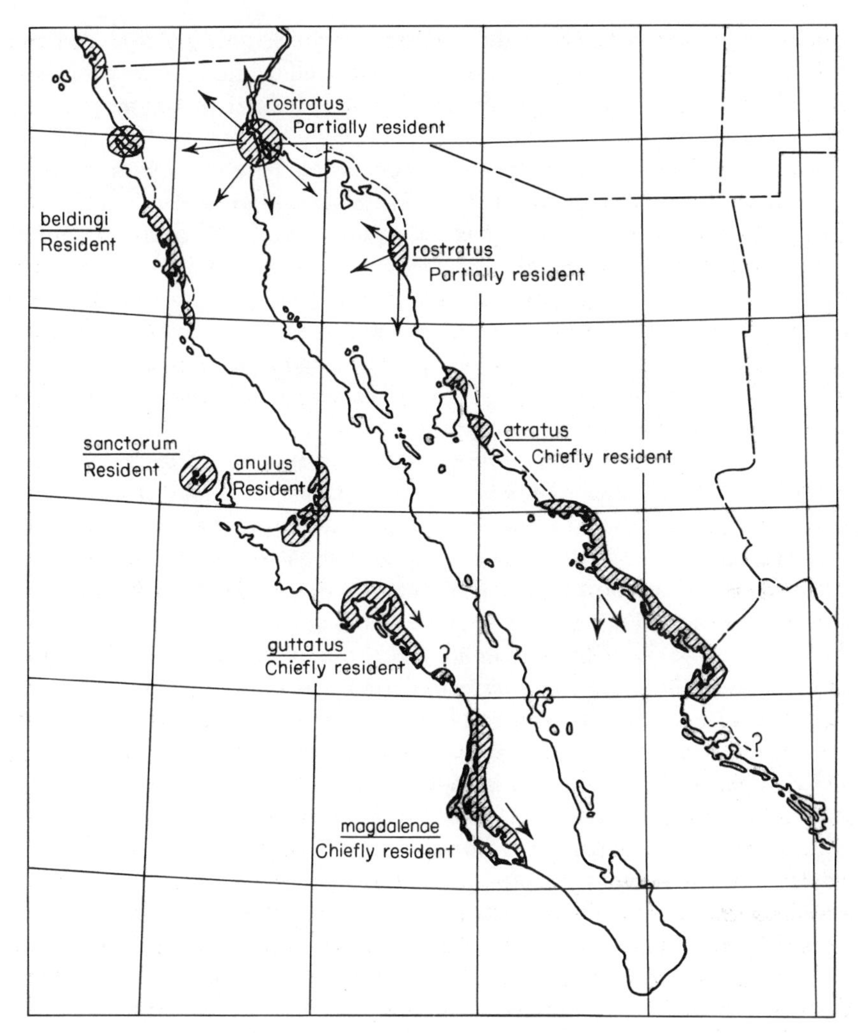

Figure 1. Distribution of the races of Savannah sparrow, *Passerculus sandwichensis,* in northwestern Mexico. (Redrawn from Van Rossem.)

ence of actual barriers to dispersal, the whole range of the species is so large that gene flow between more remote parts is prevented, and the result is an accumulation of many genetic differences in the populations of those parts. No doubt factors other than space and distance also influence the amount of interbreeding among local populations. These factors are generally ecological and may be expected to increase in effect with the increase of geographical distance. For example, the southern populations of *Rana pipiens* have a much longer period of activity each year and an earlier breeding season as well as briefer hibernation. Such factors are associated with differences in the rate of embryonic development and the time and circumstances of metamorphosis, as well as the kind and amount of food available, periods of daily activity, the kinds of predators present, and so on.

Thus the manner and kind of intergradation between extremes in a population or a series of populations is another character of high variability. The case just mentioned is an example of *clinal* variation, which occurs commonly and probably more often than not in species of fairly wide range. The broad sweep of this regular graded variation of a few characters between the extremes of the range of the species is different only in degree from that seen where one geographical race or subspecies intergrades with another along a zone in which a partial barrier has appeared but isolation is not complete.

Examples of intergradation along a chain of adjoining subspecies have been described in salamanders of the genus *Ensatina* in California (Stebbins, 1949). These subspecies are, of course, populations or groups of populations that show characters distinct enough for taxonomic recognition (Fig. 2). The point here is that such characters as the color patterns seen on the skin do not appear to show clinal differences from one end of the chain to the other but rather to indicate that the populations in particular places along the chain have undergone local modifications of their own, probably during the course of their relative isolation, because these modifications involve some striking and unusual patterns. The existence of local phenotypic evolution is more obvious than it would be among most other kinds of animals in a similar chain of subspecies. Where there is discontinuity or isolation of populations by barriers at certain places along the chain there is perhaps a reduction in a formerly wider distribution of the species. There are also examples, shown on the map, of secondary overlap between populations which can no longer interbreed even when they are sympatric (occupying the same area). Populations that are reproductively isolated when they become sympatric are usually considered to represent separate species, unless, as in this case, they are elsewhere connected by means of intervening populations among which there is genetic exchange.

Populations have two kinds of relationship to one another in an evolutionary sense. The *primary* relationship is that in which two or more populations become distinguishable but have not entirely lost contact. In the zones where contact still occurs, there is intergradation, and the degree of genetic or phenotypic differentiation between such populations is likely to be slight. This is true of most geographical subspecies, each of which consists of one or more populations. Those populations or subspecies that are adjacent to one another but interbreed along the zone of contact show less differentiation of characters than those farther apart, which may be connected by a chain of other subspecies of intermediate character.

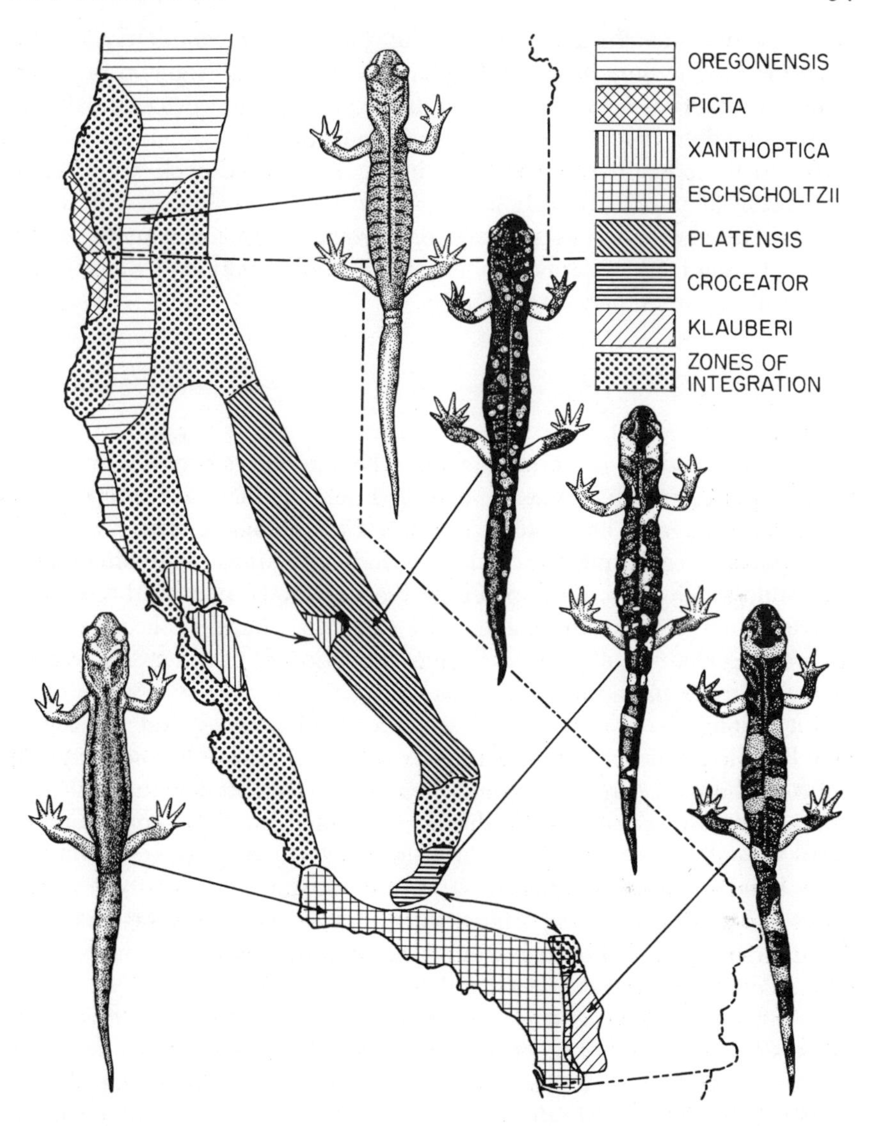

Figure 2. Distribution of salamanders of the genus *Ensatina* in California. (Redrawn from Stebbins.)

On the other hand, populations or subspecies that have been separated for some time without any intermingling and then because of expansion of range or removal of a barrier are able to come again into contact will show any of several kinds of *secondary* relationships. They may be totally incompatible in behavior and sterile in breeding, so that, although sympatric, they do not hybridize. They may hybridize in the laboratory and be more or less fertile, yet fail to do so in natural conditions. They may hybridize occasionally in nature and be partially or completely fertile. There may be a zone in which fertile hybridization is common or there may even be a complete mingling and intergrading of the two populations, obliterating their differences. The genetic contribution that one population makes to the other is referred to as introgression, and such mingling is introgressive hybridization. It may not always be possible to determine whether intergradation is primary or secondary, but frequently some circumstantial evidence from the geographical distribution of the species, or from its genetics, permits a decision.

VARIATION

Individual variation means the differences between one individual and others in the same population, insofar as these are independent of age, sex, stage in life history, or normal physiological changes. There is no point in confusing this with the kinds of changes individuals normally undergo in their development or that are shown by varied responses to stimuli in their environment (for example, changes of color in a chameleon, changes of weight during starvation or feeding, differences between larvae and adults, and so on).

But among the almost infinite number of differences that are legitimately called individual variations, it is not always possible to distinguish between those that are genetic and those that are environmental in origin, unless experiments show that during controlled, uniform conditions of development differences occur, with evidence of inheritance, or, on the other hand, among individuals known to be of the same genotype differences are produced by exposing them to varied environments. As indicated in Chapter 4, no character of an organism is due exclusively to genetic or to environmental causes.

Nevertheless, as far as environmental causes produce characteristics of organisms, these are not inherited but, obviously, only those features that can be shown to be influenced by genes, and only to the extent of that influence. Therefore only the latter are subject to evolution by means of natural selection, and the material for selection is strictly the genetic variations. Of course, favorable or unfavorable environmental effects on individuals can increase or reduce their chances of reproductive success or of survival to maturity, but it must be remembered that unless variations are transmissible to offspring, they will not contribute anything to the evolution of a species.

The causes of genetic variation have been mentioned in Chapter IV: reassortment of genes in the processes of meiosis and fertilization, recombination of genes within chromosomes by crossover, and changes in chromosome number and thus of gene number at a given locus (ploidy). These, and occasional deletions, transfers, and duplications of parts of chromosomes, provide a continual shuffling of the genome (whole genotype of a population), or, in another metaphor, a stirring of the gene pool. Natural selection has, then, in most species, a vast reservoir of possible heritable variations with which to produce well-adapted individuals or to maintain the adaptiveness of a population even if conditions of the environment vary. Mutation of genes, although rare, is apparently a universal cause of variation, in the sense that no genes are immune to it, and therefore no phenotypes can be expected to persist without its occasional influence. But the frequencies, or rates, of mutation have no necessary relation to rates of evolution, because evolution depends primarily upon natural selection, and its rate is therefore determined by such factors as selection pressure (adaptive values of variations under particular conditions) and the reproductive rate. Pleiotropic effects of genes, incidentally, may lead to unexpectedly high (or low) values for particular genes, and the additive effects of different genes on particular phenotypes can also give complex selective values to a single mutation. Thus the sources of genetic variability in any species are almost incredibly complicated.

Some of the significance of this variability in actual populations is shown in a paper by Lindsay and Vickery (1967), reporting their study of the yellow monkey flower, *Mimulus guttatus*, in the area around Great Salt Lake in Utah, formerly occupied by Lake Bonneville. The latter at its maximum covered parts of Idaho and Nevada as well as north-central Utah, a total of about 20,000 square miles. The repeated increase and decrease of Lake Bonneville in the last 50,000 years was associated with the advance and retreat of glaciers in the nearby mountains and canyons. The greater part of its basin is now available for *Mimulus* and other plants, but it is thought that the long, hot (hypsithermal) period between 7500 and 4000 years ago was unsuitable for monkey flowers in this area. The habitats now used by them probably have not been available for more than about 4000 years. On the other hand, the higher canyons may have been suitable for as long as 17,000 years, since the last general glaciation in the mountains. There is no difficulty in dispersal of *Mimulus* to either lower or higher elevations because the seeds can be carried by wind, water, and the digestive system of birds (if not mammals).

Populations of different ages (elevations) were sampled in many parts of the basin and canyons; there are hundreds of more or less separate populations in the area. Offspring of plants from twenty of these were cultivated in a standard greenhouse environment and used for detailed study of morphology, physiology, and genetics. It was shown that distinct genetic differences occurred between three quarters of the populations sampled, and one or more impediments to gene exchange in about half, although none of these were entirely isolated from the others. There had been some hybridizing with another monkey flower, *M. glabratus utahensis,* on the eastern side of the basin, with partial introgression.

> The frequency of evolutionary changes is not proportional to time. The average number of significantly different traits per population per average amount of time available are 1.75 per 12,000 years in the formerly glaciated area, 2.00 per 50,000 years for the populations of the undisturbed level, 2.25 per 14,000 years for the populations of the old lake level, and 2.50 per 8,500 (4,000 to 13,000) years for the valley populations. Furthermore, the new populations of the valleys and the glaciated areas probably first lost the distinctive traits they inherited from the population of the undisturbed level or populations from which they originated, and then evolved new ones in response to different selective pressures of their new environments. So, their real evolutionary rates are at least double the apparent ones. Greater speed is suggested by the transitory nature of some monkey-flower habitats, e.g., stream banks and sand bars.
>
> The distinctive character differences and partial crossing barriers of the individual populations apparently come and go, probably swiftly, in response to new selection pressures from their new and changed environments. The process seems to depend on a continual reshuffling of portions of the species' gene pool.

The authors believe that the differences observed between populations and the barriers to crossing that have appeared in some cases between them could have evolved in less than 4,000 years, probably much less than that. "The resulting picture is not one of steady evolutionary divergence leading to distinctive races and then species. Rather, it is a picture of a frequent emergence, change, and disappearance of distinctive populations with only the rare formation of one sufficiently distinct to be on its own evolutionary path." This careful and imaginative analysis of population genetics suggests, as some other recent work does, that divergence within species is more rapid than usually supposed, and that evolution need not proceed in a steady current involving an entire species at the same time. Also the paper shows that besides the gene flow within each local population, there is commonly a smaller flow outward, to others, and inward from others, both of these resulting in a change of the genetic composition of each gene pool, that is, serving as a further source of variability. Selection and not genetic drift seems to be responsible for the minor evolutionary changes observed.

Genetic drift is the name for purely chance fluctuations of the frequency of alleles in a population. "Drift" seems an inappropriate term unless we go back to the early recognition of this phenomenon by Wright (1931). He pointed out examples of change in both phenotypes and genotypes in small, isolated populations, that were not adaptive and could not be explained by selection, but could readily be accounted for by random reassortment of particular alleles which had little or no selective value, or, more exactly, whose selective value, if any, was overridden by accidental rearrangement. In a large population such chance fluctuations have little effect, but if a population is very small, the change produced may be greater than that brought about by selection in any given generation.

Suppose that two selectively neutral alleles exist with equal frequency in a population consisting of only twelve individuals; that is, three are homozygous, *AA;* six are heterozygous, *Aa;* three are homozygous, *aa.*

If, in the next generation, there are still only twelve individuals (or some other small number), it is not only possible but probable that one of the alleles has, by chance, gained in frequency at the expense of the other. This would be the result of *any* change in relative number. In a subsequent generation (third) it is of course possible that the pendulum would swing the other way, increasing the less common allele, but in the second generation fewer individuals were carrying it than were carrying the other, and the chance of increasing the frequency of the less common to the point where it is equally common is smaller, because of the reduced number. Likewise, the more common allele in the second generation, given the same proportional chance of increase as is given the less common allele, may thereby more probably have a large increase. This means that the less common allele has, in fact, a better prospect of becoming still further reduced than it had before. Thus, in a fourth generation of only twelve individuals the more abundant allele will probably increase and the less abundant decrease, not because either has been given a greater probable survival rate but merely because there are more of one than of the other. The chances of reduction to zero are greater for the allele that is present in lower frequency. Assume, for instance, that only three individuals carry allele *A*; the chance of complete loss, by accident, is much greater for these than it is by a similar accident for the other nine. The word "accident" in this case can include the chance of fertilization by one gamete instead of another, the discard of one set of genes in a polar body in maturation of eggs, and so on. Incidentally, the fact that one allele may be dominant and the other recessive is irrelevant here, because the phenotypes have not enough selective value to offset the effect of random change.

Thus in a few generations, assuming equal frequency to begin with, if the population remains very small the probability that the two alleles will still be present and equally frequent approaches zero because of the strong effect of random changes. Therefore very small isolated populations that happen to be cut off by barriers or distance from other populations of the same species may develop homozygosity for a number of genes (the gene is "fixed" by the loss of one or more alleles) by means of random "drift" instead of by selection. Such an effect may accompany the establishment of new colonies, the invasion of new territories such as islands, by occasional dispersal of individuals or small groups.

If, for example, an offshore island happens to be colonized by a few individuals at long intervals from a mainland population, then not only may genetic drift affect the frequency of alleles in the new island population, but there may be, as an accident of sampling, some random differences of the original founders or colonizers from the majority of the population of the mainland. The chance that one, or half a dozen, individuals represent the average of the mainland and contain alleles in exactly average proportions is not very good. This results in a modified endowment of the island population by its founders, appropriately called the *founder effect.* Of course, if large numbers of mainland individuals come at the same time or come in the course of successive seasons, or if there is frequent exchange back and forth, then neither the founder effect nor genetic drift may have any appreciable influence and the island population will not differ genetically from that of the mainland.

An example commonly cited of the effect of genetic drift in a population of few individuals is that of the blood groups A, B, and O among the Dunker sect in Pennsylvania. These people were Germans who came to the United States in the eighteenth century and founded several religious communities, virtually excluded from intermarriage with the population around them, many of whom were also of German origin. A study by Glass *et al.* (1952) showed that the frequency of alleles for A, B, and O had become very different among the Dunkers from that of either German or eastern American populations that were racially related to them:

BLOOD-GROUP ALLELE FREQUENCIES (%)

	i^A	i^B	i^O
West German	29	7	64
Eastern American	26	4	70
Dunker	38	2	60

Although drift may produce such striking effects in small, isolated populations among alleles that have little or no selective value, if one allele has a moderate selective advantage the effect of drift in a very small population is just as likely to coincide with that of selection as to work against it. If a species is divided into a great number of small populations, more or less isolated from one another, genetic drift and selection together may accomplish a particular change more rapidly than either would do by itself or than selection would do in a population not so divided.

Hardy–Weinberg Equilibrium. In a large population, on the other hand, *if* two alleles are selectively neutral, mating is random, and mutation does not occur, a particular allele tends to remain constant in frequency. This is expressed by the equation $p^2 + 2pq + q^2 = 1$, in which p is the frequency of one allele and q is the frequency of the other. No matter what the actual numbers were at a given time, they tend to occur at the same frequency in later generations provided the conditions are as specified. But, in fact, alleles are seldom neutral, mating may not be random (even if it appears to be), and mutations do, sometimes, occur. Therefore the Hardy–Weinberg equation is useful as an abstraction, against which actual frequencies may be compared, and the nature of the observed changes in a succession of generations (contrasted with the expected H–W frequencies) is an indication that selection (or sometimes nonrandom mating or mutation) is affecting the population in respect to the alleles in question. Indirectly, then, this equation is a convenient tool, but it is not a description of normal relationships in the gene pool.

Hybridization. Hybridization and *hybrid* have been applied to a wide variety of circumstances, in some of which they are not justified. A student of elementary biology is sometimes given the impression that a hybrid is the result of a cross between any individuals that differ in one or more pairs of genetic factors, and he uses the terms monohybrid, dihybrid, and trihybrid crosses. A few geneticists have done so, for instance Lotsy. But as all individuals in a mendelian population, with the exception of siblings developing from a single fertilized egg, differ in their genotypes, such use of the term would apply to all sexual reproduction and would therefore be meaningless. It is much more useful to consider hybridization as a crossing between populations that have become genetically distinct rather than the breeding of ordinarily variable individuals within a population.

This means that it is appropriate to refer to hybrids between genetically different breeds of domestic animals or varieties of cultivated plants, where differences have been brought about by artificial selection. It is also useful for crosses between members of different taxa, such as subspecies, species, or genera. The degree of taxonomic separation makes no consistent difference in the probability or frequency of hybridization. Ordinarily crossing between subspecies takes place with little or no difficulty in the areas where they meet, but there are examples of complete failure to interbreed when subspecies of one species occur sympatrically, as a result of secondary overlap of their ranges; in such cases, however, (as apparently in *Ensatina*, Stebbins, 1949), there are genetic connections between the overlapping subspecies by way of other, allopatric subspecies.

Among species, hybridization is not uncommon in nature, even though, as a rule, there occur various degrees of infertility from a slight lowering of the reproductive rate to complete failure to produce offspring. But Hovanitz (1949), among others, has shown that extensive crossing with full fertility takes place among species (or what are supposedly species) of sulphur butterflies, *Colias,* in North America, partly as a consequence of ecological changes produced by agriculture. The introduction of clover and alfalfa into newly cultivated areas can bring allopatric species together and allow them to interbreed. There is no reason to suppose, however, that this began only with human interference, for the butterflies are very active travelers. Similar hybridizing with no evidence of human influence is common in North American *Limenitis* (=*Basilarchia*) species, the admirals, and it is not at all clear where the lines can be drawn between species, subspecies, and hybrids among these nymphalid butterflies.

Numerous examples of natural hybrids are found in freshwater fishes, involving some 175 species combinations in North America and 75 in Europe, and there are at least 11 among marine fishes (G. R. Smith, personal communication). This subject is discussed most fully by Hubbs (1955, and other papers). Many cases can be attributed to human introduction of freshwater fishes into regions where they did not previously occur. Hybridization is rare between species of mammals but moderately common in birds (see, for instance, Mayr, 1963, p. 126, and references therein). In plants it is very common, and because chromosomal differences between species are sometimes rendered harmless by polyploidy, allowing continued reproduction of hybrids between species in nature, many plant species can be traced to a crossing of parental species; this is, of course, one of the commonest ways of producing new "varieties" of cultivated plants, but it does not work in animals.

There is a curious case among certain salamanders, in which triploid females occur and are maintained in the population along with diploid females and males. *Ambystoma jeffersonianum* occurs in the north-central states and a part of the northeast; *A. laterale* is a smaller, more northern species. Associated with both of these are triploid females, different enough to have been described as species, *A. platineum* and *A. tremblayi,* respectively. (See Uzzell, 1964, and MacGregor and Uzzell, 1964.) The monoploid number of chromosomes is 14; *jeffersonianum* and *laterale* are diploid (28), and the associated triploid females have 42. These females are not reproductively isolated, but their eggs are activated by sperm from the corresponding diploid males. The sperm nuclei, however, do not contribute to the nuclei of the eggs but remain in the cytoplasm, and therefore males have no genetic effect upon the offspring, which are again triploid females. This is a case of gynogenesis, not parthenogenesis.

According to Uzzell (1964) there is a precedent for recognizing the triploids taxonomically, by a specific name, as in the small fish *Poecilia formosa* Girard. (Hubbs and Hubbs, 1932). These are all females and the "species" is thought to have arisen by hybridization of two other species of *Poecilia.* These triploid females mate with the males of two species but produce only triploid females. It has been suggested that in such a case, if there is an advantage to these females they should eventually replace the diploids. But diploid males can only be produced by diploid females, so the population could not be reduced too far because males are necessary for activating the eggs of either. Although the triploid females are recognizably different, they are not reproductively isolated and are able to reproduce only because of the presence of males. It therefore seems that they cannot be considered species in any biological sense.

Polymorphism. Long before the genetic basis of variability was understood, many species of animals and plants were known to exist in more than one "form" or to show distinct differences in certain characters, with or without intermediates. Such species were called dimorphic, trimorphic, or polymorphic. Darwin, discussing individual differences in *The Origin of Species*, thought it "a highly remarkable fact that the same female butterfly should have the power of producing at the same time three distinct female forms and a male," and he said: "These facts are very perplexing, for they seem to show that this kind of variability is independent of the conditions of life."

They were not only perplexing but a cause of considerable difficulty to taxonomists who were attempting to classify species and genera in which such variation occurred. In particular, it was something of a challenge to the idea of fixity of species, if not of their reality. Many of these questions have now been answered, and the current concept of species rests largely on the ability of individuals to interbreed, and the possession of a common gene pool by a population that nevertheless remains apart from other populations because of barriers to genetic exchange. Yet the range of variation may be great and even discontinuous in respect to some characters. Generally polymorphism is due to the presence of alleles at one or more loci; these have arisen by mutation in the first place and have been sustained or even promoted by selection in many cases. In other cases they have remained in the population because of some selective advantage incurred by heterozygous individuals.

Gradually, as the study of genetic variability increased, any kind of polymorphism came to be represented in the mind of the geneticist by its known genetic basis, that is, by the pair or pairs of alleles involved, or sometimes by chromosomal abnormalities that had a similar effect. Thus chromosomal polymorphism and genotypic polymorphism were seen as the sources of many individual differences. Today an evolutionary geneticist uses the word "polymorphism" to mean any genetic differences caused by the continued presence of alleles. To the extent that a population is heterozygous, it is polymorphic. All genetic differences constitute polymorphism if they are maintained in a population beyond the frequency of occasional mutants. This seems unfortunate, for the word no longer has any special meaning apart from ordinary genetic variation. What once seemed to be an exceptional and interesting kind of individual variation has been blended with other degrees of variation until its distinctiveness was lost, and it became part of a nearly universal phenomenon.

In nongenetic terms, if polymorphism is to retain any useful meaning, it ought to refer to those sustained variations that are discontinuous, as between blood groups, eye colors, and contrasting patterns (as the variants of *Rana pipiens* [Volpe, 1961], cinnamon and black bears, and so on). In moths that illustrate the selective shift to "industrial melanism" their status in regard to color has been called "transient polymorphism," assuming that the frequency of melanic individuals has not yet become stable.

REFERENCES

Emlen, J. T., Jr., and G. B. Schaller. 1960. Distribution and status of the mountain gorilla. *Zoologica 45:* 41–52.

Glass, H. B., M. S. Sacks, E. F. Jahn, and C. Hess. 1952. Genetic drift in a religious isolate: an analysis of the causes of variation in blood group and other gene frequencies in a small population. *Am. Naturalist 86:* 145–160.

Hovanitz, W. 1949. Increased variability in populations following natural hybridization, pp. 339–355. In G. L. Jepsen, G. G. Simpson, and E. Mayr (eds.), *Genetics, Paleontology and Evolution.* Princeton Univ. Press, Princeton, N.J.

Hubbs, C. L. 1955. Hybridization between fish species in nature. *Syst. Zoology 4* (1): 1–20.

———, and L. C. Hubbs. 1932. Apparent parthenogensis in nature, in a form of fish of hybrid origin. *Science 76:* 628–630.

Lindsay, D. W., and R. K. Vickery. 1967. Comparative evolution in *Mimulus guttatus* of the Bonneville Basin. *Evolution 21:* 439–456.

MacGregor, H. C., and T. M. Uzzell. 1964. Gynogenesis in salamanders related to *Ambystoma jeffersonianum. Science 143:* 1043–1045.

Mayr, E. 1963. *Animal Species and Evolution.* Harvard Univ. Press, Cambridge, Mass.

Miller, R. R. 1961. Man and the changing fish fauna of the American southwest. *Papers Mich. Acad. Sci., Arts, Letters 46:* 365–404.

Moore, J. A. 1949. Geographic variation of adaptive characters in *Rana pipiens* Schreber. *Evolution 3:* 1–24.

———. 1950. Further studies on *Rana pipiens* racial hybrids. *Am. Naturalist 84:* 247–254.

Ruibal, R. 1955. A study of altitudinal races in *Rana pipiens. Evolution 9:* 322–338.

Stebbins, R. C. 1949. Speciation in salamanders of the plethodontid genus *Ensatina*. *Univ. Calif. Publ. Zool. 48:* 377–526.

Uzzell, T. M. 1964. Relations of the diploid and triploid species of the *Ambystoma jeffersonianum* complex (Amphibia, Caudata). *Copeia 1964* (2): 257–300.

Volpe, E. P. 1961. Polymorphism in Anuran populations. In W. F. Blair (ed.), *Vertebrate Speciation*. Univ. Texas Press, Austin.

Wright, S. 1931. Statistical theory of evolution. *Am. Statistical J.* (March suppl.): 201–208.

6 · SPECIATION

The origin of species is a complex process involving much of what has been discussed already, plus the separation of populations so that they do not interbreed with each other. As seen in the case of *Mimulus* in the Great Basin, many species consist of populations more or less isolated, more or less differentiated genetically, and more or less adapted to local conditions. But the question, "When does such a collection of populations become two or more species?" has little meaning and is almost unanswerable. It does not depend upon the degree of difference established, for if a barrier arose that persisted for an unlimited time so that there was no further connection between the gene pools separated by it, then by the "biological" criterion species were established when the barrier became effective, assuming that man was not present to put them to an experimental test. Nor does the question depend upon the accumulation of enough genetic differences to provide sterility if the separated populations again became sympatric, for, as already noted, there are a great number of examples from nature of species capable of hybridizing. It does not depend upon establishment of phenotypic differences useful to a taxonomist, because a considerable number of "sibling species" are known in which there are no such differences, yet populations are separated by a sterility barrier, or a courtship or mating barrier, however it may be achieved. We can hardly even say that our recognition of species depends upon the judgment of a competent taxonomist who is in possession of the relevant facts, for there are valid differences of opinion based upon the same evidence.

Apparently, then, there is no rule that will determine when one species becomes two, that is, when two species are produced by the fission of one population. The distinction between them, no matter how obvious it eventually becomes, may for a long time be quite intangible. Biologists are well agreed that reproductive isolation of some kind is the essential element of speciation. In a few instances, as the hybridizing of species of plants, resulting in fertile offspring with a different chromosome number, unable to cross with either parent for that reason, we can say that a "new species" arose at one step, with reproductive but not geographical or ecological separation. Otherwise, biologists favor actual geographical isolation as the probable means by which populations are given the opportunity to develop their own characters independently (and thus differently). From then on there is not just a random accumulation of differences (although this would be possible by means of genetic drift in very small populations) but rather a selective response to different and changing circumstances. Only in a few cases does it seem as if speciation might possibly have occurred without geographical isolation, but these are open to more investigation than they have received.

SPECIES FLOCKS

One of the most interesting of these is the occurrence of "species flocks" in some of the oldest lakes, notably Baikal, Tanganyika, and Nyasa. A species flock is a group of closely related species that have evolved from, presumably, one ancestral stock within the lake. Many such flocks or swarms are known, the most fully investigated being among the amphipod crustaceans and cottoid fishes of Lake Baikal in Siberia and the cichlid fishes of the great Rift Valley lakes, Tanganyika and Nyasa, in Africa. At first it would seem impossible that a population of fishes occupying a single body of fresh water could, without some unusual evolutionary mechanism, become segregated completely enough to give rise to scores (in one case more than 100) of descendant species of many genera. But the situation is not actually so simple. These lakes are of great size and ecological diversity, and they have had a complex history, all being in regions of active geological change.

Brooks (1950; see for further sources) reviewed comprehensively the work so far done on these and other ancient lakes in reference to species flocks, the total fauna, ecological data, the age and history of each lake so far as known, and the problem of speciation. A brief résumé of the pertinent information follows.

Lake Baikal is in a long, deep, intermontane trough in southern Siberia, latitude 52–56°N. Its length is 675 kilometers, width about 50 km, the surface at 463 meters, greatest depth 1741 m. The bottom temperature is about 4°C, but unlike other deep lakes, oxygen is available throughout. It is the deepest and oldest lake in the world, dating from at least the Paleocene, perhaps the late Cretaceous. The outlet is by the Angara River at its southern end, to the Yenesei River and the Arctic Ocean. The fauna has been studied extensively, and is all, with two exceptions, derived from widespread, well-established freshwater groups. The exceptions are the Baikal seal, *Phoca sibirica,* and an amphipod, *Gammaracanthus* sp. Both of these are only subspecifically different from their nearest relatives in the Arctic Ocean and must be very recent arrivals in the lake, perhaps late Pleistocene, when there was an extension of the sea up the valley of the Yenesei.

Otherwise, much of the fauna consists of endemic species (those that originated in their present location), genera, and families. The amphipods have been studied most thoroughly as to systematics and ecology; there are 290 endemic species in 30 genera, 27 of the latter being endemic. There were probably 18 original stocks from which species flocks originated. Almost all are benthic in habit, which means that their distribution in the lake is related to and dependent on ecological features of the bottom, some of which are favorable, and others form barriers to dispersal. Of ostracods, an order of minute crustaceans, there are 36 endemic species in 3 genera, the latter also occurring elsewhere. Among fishes, besides 12 widespread Palearctic species there is an endemic group of 18 species of sculpins in 10 genera. Two species belong to *Cottus*, a well-known genus of the northern hemisphere (family Cottidae); the others, for which *Cottus* must have been the original stock, are in 8 genera of the family Cottocomephoridae, benthic and pelagic fishes that all lay benthic eggs, protected by the males, and 1 genus of another family, Comephoridae, with 2 species that are pelagic and viviparous, but differ in their breeding seasons. The latter 2 families occur only in Lake Baikal. The benthic habit of many of these, coupled with the care of eggs by the males, would make it probable that in a very large lake there is geographical segregation of populations while breeding, if not at other times. The partial division of Lake Baikal into two basins has an isolating effect on some of the deep-water animals, and both the extent and depth of the lake have undoubtedly changed through its history, so that it has at times been partially or wholly divided.

Lake Tanganyika is between 3° and 9°S, length 650 km, width about 50 km, maximum depth 1435 m (second deepest in the world). Below 300 m the temperature is 23.1°C, and there is no oxygen below 100 m. In summer the upper 60 m reaches 25.75°; a slight cooling in "winter" allows partial mixing of the water to 100 m, but oxygen is very slight below 60 m. There is not enough clear evidence on the age of the lake, but it may be Miocene, probably older than Nyasa. The outlet is by the Lukuga River to the Congo River, and the fauna is related to that of the Congo.

Lake Nyasa is of similar shape, as both lakes occupy parts of the Great Rift Valley. It is 580 km long, from about 9 to 15°S, and has a maximum depth of 706 m, the fourth deepest lake. Below 300 m the temperature is 22°C, and there is no oxygen below 100 m. The thermocline, a zone where the temperature of the upper levels changes noticeably to that of the deepest part, is distinct but not as abrupt as in Tanganyika. The age has been considered Pleistocene but may well be Pliocene. The Rift Valley was formed gradually, not all at once.

The fish faunas of these two lakes are remarkable for their flocks of endemic species, especially among the Cichlidae, a widespread family of perchlike fishes. A tabulation of the numbers is as follows:

	LAKE TANGANYIKA	LAKE NYASA
All fishes		
Families	18	13
Genera	79	44
Species	162	223
Cichlids		
Genera	34	20
Species	91	178

"The cichlids have speciated out of all proportion to the representatives of other families. There are 174 endemic cichlids in Lake Nyasa, all but 5 of which are related to *Haplochromis* . . . 101 are placed in this genus. The more divergent . . . in 20 endemic genera" (Brooks, 1950). Trewavas (1948), who analyzed these and other species flocks, considered that multiple invasion from outlying rivers and other sources is neither necessary nor possible as an explanation for the amount of speciation here. Brooks thinks that these groups have evolved by geographical speciation within each lake, by isolation in different benthic and littoral environments in moderately shallow water. (There would be insufficient oxygen below 60 m or less, and the cichlids have specialized breeding habits, as any tropical-fish fancier knows, which tend, at breeding times, to restrict the distribution of their populations.)

There is some information on the biology of *Tilapia,* for instance, of which 4 species form a closely related group in Nyasa. Two of the species, locally called Chambo and Lidole, "are phytoplankton feeders and differences in feeding habits are difficult to detect, although some divergence may exist, as Chambo feeds closer to shore than does Lidole. . . Chambo carries the eggs and shelters the young until they have reached a total length of 15 mm., at which time she deposits them in a reed bed. Lidole continues to shelter her young until they have reached a total length of 40 mm., dispensing with the reed beds. The mating areas of these two species, whose feeding areas overlap, are separate. . ." (Brooks, 1950). Under such circumstances, when varied local environments are available, it is not difficult to see that certain requirements of a species limit its breeding distribution and may produce partial or complete isolation of one population from another. The geographical pattern of bays, shallows, deeps, ridges, wave-beaten or quiet coast, rocky or muddy shore, and so on, is changeable both by tectonic movements (uplift or sinking of the earth's crust) and by the normal erosion–deposition cycle around the lake. Both physical and biological factors must have played a part in bringing about this rapid speciation.

Another point of interest is that relatively few barriers are actually required, either as a succession in time or contemporaneously, to accomplish a large amount of speciation. To picture the simplest possible case, imagine a body of water containing one widely distributed population, then divided by a single barrier that persists long enough to allow genetic differentiation to occur before the separated populations come together again. Let the barrier be removed and the two mingle as species that do not interbreed. Another barrier, if formed in the same way (not necessarily the same barrier, or location), can now produce 4 species from the 2 species that have mingled. Assuming only that this event happens, the barrier persisting each time long enough to allow divergence, then the populations remingling (and survival of the stocks in question), it will be seen that the number of species so made would increase by geometric progression: 2, 4, 8, 16, 32, 64, . . . (the last after six successive barriers). Naturally, like most such "models," this is merely suggestive and does not represent a real instance. But it shows that we do not have to imagine some separate, discrete barrier to bring about *each* case of speciation. There has been ample time and space, and probably sufficient topographic change, to make it highly probable that speciation in large lakes has followed the same principles as that on land. (See also Myers, 1960, on the rapid evolution of fishes in Lake Lanao.)

SPECIATION OF INSECTS

Now we may look at another problem: What factors may have contributed to the huge diversity of species in the class Insecta, which outnumbers, in species, all other organisms put together? This is one of the most impressive of all biological phenomena. More than three fourths of all species of animals are insects, and the number now described stands close to a million; some 10,000 are added each year. Incidentally, about one out of every four species of animals is a beetle.

Rather than generalize broadly about insects as a whole, it may be well to look first at some qualities of the most primitive insects. Their system of air tubes for respiration, and an external skeleton moved by internal muscles, limit them in several ways. First, in size, for it seems that a respiratory system of this kind, with accompanying physiological features such as the hemolymph, could not be sustained in an animal as large as a rabbit, for example, nor would these features lend themselves to the further adaptation of a high, sustained metabolic rate. In spite of the remarkable facility of muscles and joints in an animal of minute size, the mechanical requirements of strength and mass of skeleton in an animal weighing several pounds would exclude the possibility of flight and probably of agility in running or jumping, if the skeleton were external.

The primitive insects, unlike spiders, were not committed to a predatory habit but were probably saprophragous and consumed dead vegetation as it stood or where it fell. The primitively wingless orders Protura, Thysanura, and Collembola retain these habits even now, because the niche available for them extends over most of the land surface of the earth and is only partially occupied by other groups of animals. But obviously there are other habits related to these, to which transition might easily be made. One is to feed on various parts of living plants before they die, and probably this is characteristic of more than half of the kinds of insects. Another is to feed on fungi, which themselves are either parasitic or saprophytic. A great number of insects in many orders are fungus feeders. The further division of insect feeding habits into those that are predatory and those that are parasitic upon other insects has certainly occurred many times in various orders, but the principal question here is the amount of speciation occurring in insects as compared with other groups.

The development of the plant kingdom on land resulted in creation of an enormous diversity of microhabitats, all of which were open to exploitation by animals that were small enough and adapted to feeding on vegetation. There are, of course, reciprocal adaptations of plants and insects, not only in the well-studied example of insect pollination of flowers, but in the ability of trees to replenish their leaves when stripped by insects, or in the ability of herbaceous plants such as grass to continue to grow and replace that which is destroyed. Most of the higher plants furnish not only an external surface but several internal environments, as between the layers of leaves, between the bark and the wood, within the wood, within petioles or needles, and within the tissues of galls that are induced by the presence of insects. Also there are niches on the surface of bark, on the surface and interior of roots, inside seeds or fruit, flowers, and so on, all of which are used by many kinds of insects.

In addition to this, the vegetation that covers any particular area of the land creates its own physical environment to a large extent by conserving water, reducing the rate of evaporation from the surface of the ground, maintaining atmospheric humidity close to the ground, creating shade and interrupting the wind, contributing organic matter to the soil, and creating a variety of surfaces and spaces that are open to animals. Fresh water, especially the smaller bodies, does not exclude insects, for it is usually well supplied with organic matter, and insects can either carry with them their own air from the surface or develop gills that are integrated with the tracheal system. On the other hand, very few of them are equipped to face the quite different environment of the ocean.

The ability of insects to evolve in terrestrial habitats is an example of *opportunism* in evolution, for in most of these habitats, in the soil or in dead or living vegetation, they have few competitors except among themselves. Reciprocal adaptation has probably been an impetus for development of both the herbivorous types and their predators and parasites. It is not clear that the factors facilitating speciation differ in any fundamental way in insects from those that obtain among other animals, but the *opportunities* to adapt and speciate are evidently much greater, especially in the rich and favorable environments produced by vegetation. The tropical rain forest shows this in the highest degree.

REPRODUCTIVE ISOLATION

Numerous authors have discussed in recent years the ways in which one population is prevented from interbreeding with another closely related to it, at a time when the two may recently have begun to diverge. Almost invariably the discussion is related to the process of speciation rather than adaptation, because some kind of separation of gene pools is essential to divergence of species. The exchange of genes would in many cases have a swamping effect, preventing the formation of different genomes (population genotypes). But meanwhile natural selection continues to act on each population according to the circumstances it meets and the kinds of variation that occur in it. These effects are not always different, but it is probable that environmental factors and the available genotypic variations will differ enough in various areas occupied by the original population so that if separation is achieved, there will be a gradual accumulation of genetic characters distinguishing one daughter population from another.

For instance, Stebbins (1950) described cases of seasonal isolation between species of pines in California, which are closely related and mutually fertile but produce pollen at different times. The Bishop pine is a northern coastal species; the Knobcone pine an interior, montane species; and the Monterey pine is native to the central and southern coast; but in the vicinity of Monterey their ranges overlap. Each species has its own, easily recognizable characters (especially of the cones), plus a narrowly limited season in which the female cones are receptive to pollen, usually only a few days. The habitats differ somewhat, and the Monterey pine is pollinated in February, the other two in April. As a result there is very little hybridizing. Evidently the differences, including those of season, evolved while these species were wholly separated and reflect, in part, climatic adaptation. But now in the area of sympatry (geographical overlap) some of the adaptations are such as to interfere with successful reproduction between the three populations. (Many other examples are cited by Grant, 1963.)

Thus there are two important aspects to the accomplishment of speciation: (1) the actual physical separation by barriers, which usually are ecological in effect, as well as geographical; and (2) the accumulation of adaptations to the local environment of each population, resulting in genetic divergence. This divergence frequently, but not invariably, entails differences that will serve to prevent interbreeding of the two populations if later they occupy the same area.

If we list the ways in which one population can be reproductively isolated from another, they include separation in space, time (as of breeding), and ecological situation; mechanical, behavioral, and genetic incompatibility (the latter causing defects of development even if fertilization is accomplished); and various degrees of hybrid inviability or sterility. This list includes a number of factors that are not comparable with each other in any logical way, having nothing in common except that they may, sometimes, be obstacles to interbreeding of populations. Some of the factors prevent contact between members of one population and those of the other, at least at the time of reproduction. Other factors in the list do not prevent contact but interfere with reproductive success. With no more information than this, it would seem clear that none of these impediments to interbreeding is necessarily the result of an adaptation by means of natural selection to prevent reproduction, unless we should find that it has no other possible explanation. Some biologists agree with Moore (1957) that the various divergent adaptations of separated populations "involve differences that will also serve as isolating mechanisms. There is no critical evidence that isolating mechanisms develop as *ad hoc* contrivances that prevent hybridization between incipient species. Furthermore, there are some theoretical considerations that make it improbable that the genetic differences that serve as isolating mechanisms commonly arise in this manner."

The term "isolating mechanism" is used by some authors currently to designate all the ways in which reproduction among populations is prevented, whether these are merely separation by distance, ecological barriers, or structural and functional differences in the organisms themselves. Others limit the meaning of isolating mechanisms to those characters in the organisms that have some possible relationship to breeding or reproduction. The idea is widespread that these characters are mechanisms developed by natural selection for the purpose of producing reproductive isolation. Stebbins (1966) says, for example, that the processes which lead to the origin of species "consist of the divergent action of natural selection upon those particular kinds of differences in genes and chromosomes which can form isolating mechanisms." This seems to imply that such a characteristic as the difference in time of pollination in two closely related species of pine trees that occur in the same area was brought about by natural selection, and maintained, in order to prevent interbreeding between them. In this case it appears that the original adaptive value of an early season of pollination (suited to a warmer climate) and a later season (appropriate to a colder climate, with shorter summer) has been forgotten, or, on the other hand, that natural selection is supposed somehow to have anticipated the possible future overlapping of two populations that were separate at the time the adaptation arose. The latter is quite out of the question, of course, because natural selection acts only upon characters that exist and in circumstances that affect reproductive success; it is not capable of anticipating anything, for there is no provision in nature for completely bypassing cause and effect.

If we grant this and allow no farsighted demon to control selection, what is the possible advantage in preventing breeding between two such closely related species? Lists of such advantages commonly include: preserving the integrity of each gene pool, maintaining the adaptive homogeneity of each population, preventing the futile production of hybrids that are at a disadvantage, eliminating "wastage of gametes," and so on. An additional question must then be answered: How can selection possibly bring this about? Presumably it must build up in the successful reproducers (those that did not waste gametes, produce hybrids, breed outside the gene pool from which they came, and so on) a set of mechanical, behavioral, physiological, or simply genetic features that ensures more and more completely against cases of mistaken identity in mating or mistaken acceptance of the wrong pollen grain by a female cone or stigma. Any means by which this is accomplished, *for this function primarily*, so far eludes us.

Looking a bit further into actual cases, it is virtually impossible that a male *Rana pipiens* living in Vermont will, under natural conditions, mate with a female of the same species in Texas, or a male cankerworm moth emerging in the spring with a female that emerges in late fall, or plethodontid salamanders that breed in water with others that spend their lives normally under logs and lay their eggs there. Were these seasons, regions, or ecological niches elected by separated populations in the first place so that the purity of their gene pools might be sustained? Put in these terms the notion is absurd, for these represent the results of dispersal in space, time, and ecology for adaptive reasons that had nothing whatever to do with reproductive isolation. Let us suppose that because of differences in size, structure, or behavior it is difficult or impossible for mating to take place between two related species of birds that occur in the same area. Such differences may arise fairly rapidly in evolution, although no more so than most other differences. When they do occur they are correlated with other adaptive differences between the species in question. Often there seem to be no obstacles that we can observe, yet members of the two species do not mate with each other. Such a situation occurs with the two species of boat-tailed grackle, *Cassidix*, where they overlap along the Gulf coast of Texas; it has been found that these virtually identical "sibling" species are distinct, and remain so, because the females will respond to the courtship of males of their own species but not to the slightly different courtship of the other. There is no reason to doubt that these differences were established during a period of separation of two original populations, and serve, in the present area of sympatry, as a means by which females respond to conspecific males.

A swamp pool in the spring may contain frogs of several kinds, perhaps some of them being closely related species of *Pseudacris*, as in the southeastern states. No mechanical difficulty in mating exists, but there is a behavioral problem. In the chorus of males, the voices of *P. nigrita*, *P. brimleyi*, and *P. triseriata*, for instance, differ in pitch, pulse (trill) rate, and duration, and are therefore quite easily distinguishable. It has been shown that females respond to voices of males of their own species by approaching them. If a male attempts amplexus with a female not of his species he is not accepted but shrugged off as a nuisance. Should we regard the refusal of a female to accept a male of a different species or to react to his voice as isolating mechanisms developed by natural selection to guard the integrity of the *P. nigrita* gene pool, or simply as a consequence of the fact that she is only disposed to mate with one of her own species? Adaptations are continually maintained in a population to facilitate its reproduction, and one of these adaptations is the recognition and acceptance of one sex by the other.

Students of isolating mechanisms (for example, Mecham, 1961), however, urge an additional bit of evidence that reproductive characters are, first of all, *isolating* mechanisms: In an area of overlap between two closely related species the difference in calls of the males is intensified, thereby increasing the probability that a female will keep away from males not of its own species, whereas in areas where the species are not sympatric the calls do not differ so much. In the midst of a noisy chorus of many frogs, is a female required to decide first that so-and-so is not for her? It would seem much simpler merely that she recognize a distinctive voice, the only voice that causes her to react, namely that of her own kind. All others are mere noise, for which no attention is necessary and no decision called for. The enhanced frequency of union of *nigrita* with *nigrita* and *triseriata* with *triseriata* seems a sufficient reason for intensifying the difference of voices where there exists some possibility of confusion. The failure of any pairs to produce progeny because of mistaken identity is merely the reverse side of the same coin.

Field and laboratory studies of "isolating mechanisms" are now very numerous. They are undertaken most frequently to find examples of characters that can be supposed to have originated through natural selection to prevent mismating, or hybridizing. But as the method by which selection can do just this has not been demonstrated, it seems to the present writer that perhaps it is, after all, simply that most species of animals and plants are adapted in quite intricate ways to reproduce their own kind successfully. The complexity of the whole process of courtship, breeding, fertilization, embryonic development, and, sometimes, parental protection, is great enough so that few, and minor, differences at any stage reduce its success or cancel it entirely.

Is there any critical evidence from experiments that might be brought to bear on production of isolating mechanisms by selection? A paper by Koopman (1950) is often cited. Two species of *Drosophila, pseudoobscura* and *persimilis,* capable of hybridizing, were maintained together in cages for many generations. They were "marked" by genetic characters that would show in hybrids, and in each generation these hybrid individuals were removed. In the course of a number of generations the frequency of hybridizing was brought down practically to zero, showing that selection clearly had an effect. This result has been widely but not universally interpreted to mean that selection has brought about the formation of genotypes whose function is to prevent crossing between these two species. But what these genes may be or how they act is difficult to imagine. There already exist, as an important part of the genetic makeup of any population, provisions to promote mating and reproduction between individuals that respond to each other because they are members of the same species. Such adaptations are reciprocal between one sex and the other. The elimination of hybrids in the experiment has simply reduced the variability of such characters in each of these species so that there is even less capacity than before for responding, occasionally, to members of the other species, and a stock is left that breeds only with its own species.

REFERENCES

Brooks, J. L. 1950. Speciation in ancient lakes. *Quart. Rev. Biol. 25:* 30–60, 131–176.

Grant, V. 1963. *The Origin of Adaptations.* Columbia Univ. Press, New York.

Koopman, K. F. 1950. Natural selection for reproductive isolation between *Drosophila pseudoobscura* and *Drosophila persimilis. Evolution 4:* 135–148.

Mecham, J. S. 1961. Isolating mechanisms in Anuran amphibians, pp. 24–61. In W. F. Blair (ed.), *Vertebrate Speciation.* Univ. Texas Press, Austin.

Moore, J. A. 1957. An embryologist's view of the species concept. In E. Mayr (ed.), *The Species Problem.* Am. Assoc. Adv. Sci., Publ. 50, pp. 325–338.

Myers, G. S. 1960. The endemic fish fauna of Lake Lanao, and the evolution of higher taxonomic categories. *Evolution 14*(3): 323–333.

Stebbins, G. L. 1950. *Variation and Evolution in Plants.* Columbia Univ. Press, New York.

———. 1966. *Process of Organic Evolution.* Prentice-Hall, Englewood Cliffs, N.J.

Trewavas, E. 1948. The origin of the cichlid fishes of the Great African Lakes, with special reference to Lake Nyasa. *Abstract Sect. V. B., 13th Congr. Intern. Zool., Paris.*

7 · ADAPTATION

MEANINGS OF ADAPTATION

As a *process*, adaptation is not quite synonymous with evolution, but is the most obvious aspect of it. Adaptation is the production and maintenance in a population of characteristics that fit it for a particular way of life (whether changing or not). Adaptation is brought about by natural selection of heritable variations, but this has two forms: selection directs adaptive evolution, and it prunes out maladaptive variation.

The word "adaptation" is also used for the *results of the process*, whether structural, physiological, behavioral, or any combination of these. As virtually every characteristic of every organism takes part in fitting it for its normal manner of living, one may understandably refer to an animal or plant as a "bundle of adaptations" (Fig. 3). But this is speaking loosely, for the organism as a whole survives, reproduces, or dies, not its parts; it succeeds or fails as a whole. Only for convenience do we discuss adaptive characteristics separately, as in following parts of this chapter.

A third meaning of adaptation is the *changes made by an individual* during its life, by which it adjusts to varying conditions internally or externally. These changes obviously could include the whole range of self-regulatory processes that are found in any organisms, but the present brief discussion may be simplified by noting three common categories:

1. Physiological adaptations brought about by reflex action, for which certain internal or external changes serve as stimuli. The pupillary reflexes, for instance, are contraction of the circular muscle of the iris when the entering light is bright, or its relaxation (dilation of the pupil) when the light is dim. In response to reduced partial pressure of atmospheric oxygen, as at high altitude, the smooth muscle fibers in the capsule of the spleen contract, forcing a release of red blood cells to the general circulation that were previously accumulated in "dead spaces" within the spleen; this permits increased oxygen transport by the blood. In response to high arterial pressure in the carotid body (in mammals, at least), there is a reflex inhibition of the output of the heart, and, on the contrary, high

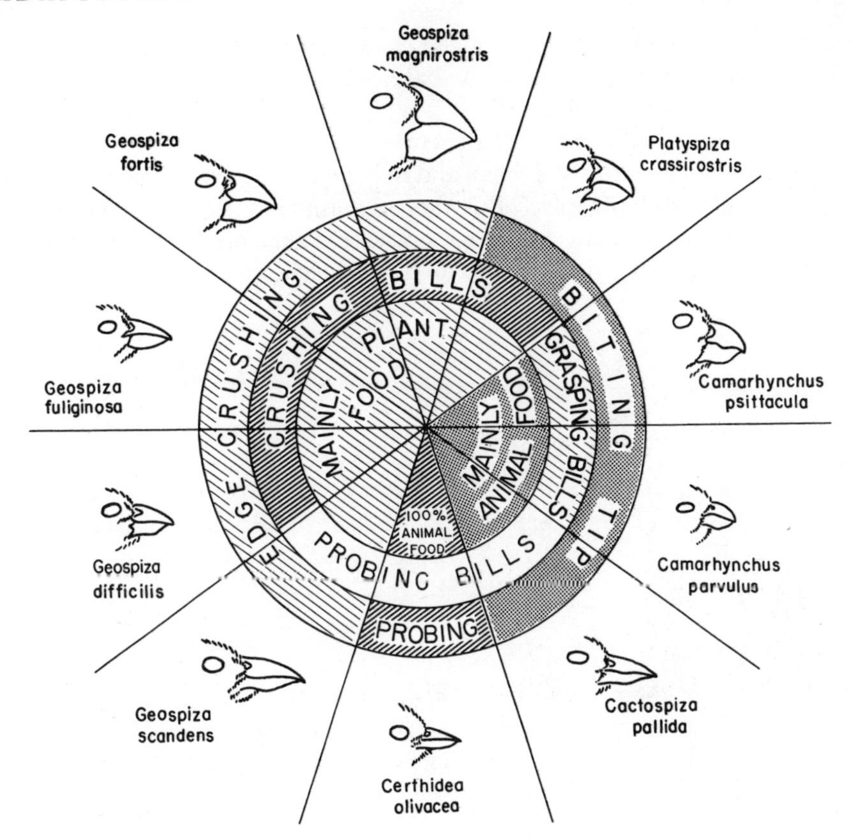

Figure 3. Morphological and functional differentiation of the bills of the Galápagos finches. (Adapted from Bowman.)

pressure within the right auricle causes, reflexly, increased output. Thus all these mechanisms act to restore to equilibrium an activity that has been temporarily out of balance.

2. Changes in structure and function induced during development by different environmental circumstances, as, for example, the variation of growth rate and degree of differentiation in seedlings according to intensity of light, those grown in normal daylight developing their leaves and chlorophyl while relatively short, but those in darkness being taller and less differentiated (without chlorophyl or leaves) (Figs. 4 to 8). Again, some kinds of semiaquatic plants, such as the mermaid weed, *Proserpinaca,* grow with some of their leaves in the water and finely divided, but with others above the surface, narrow and serrate; this difference is abrupt at the point where the waterline happens to come.

3. Seasonal changes of structure or function in an individual, for instance the change of plumage and behavior of many birds between breed-

ing and nonbreeding seasons, or the change of pelage of arctic mammals such as the ermine and snowshoe hare, which are brown in summer but white in the long winter.

These individual changes of structure or function are adaptive, no matter what stimulus provokes them, and may be called, with some ambiguity, "physiological adaptations." But the organism has simply inherited the capacity to make them, under circumstances that are met from time to time during its life; the ability to do this is, itself, an evolutionary adapta-

Figure 4. Differences in stems and leaves of bean seedlings grown in light of different intensities: (A) 100 percent; (B) 50 percent; (C) no light.

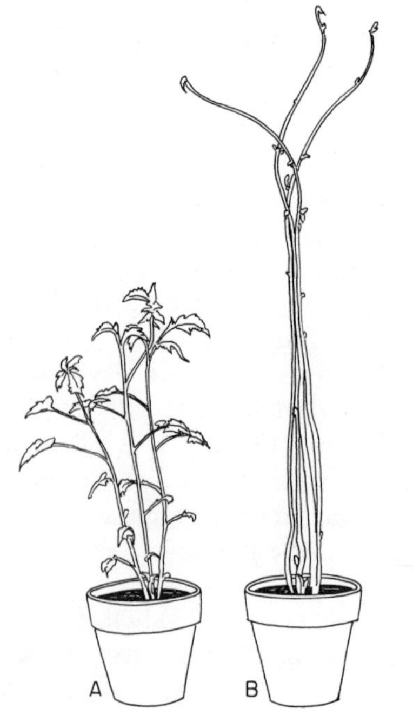

Figure 5. Potato sprouts that grew in light of different intensities: (A) in greenhouse; (B) in darkroom.

Figure 6. Mermaid weed, *Proserpinaca,* showing difference between leaves that are submerged (divided) and those that are out of water (not divided). (Redrawn from E. N. Transeau, H. C. Sampson, and L. H. Tiffany, *Textbook of Botany,* 1953, courtesy of Harper & Row, Publishers.)

tion of the population or the species. It is important to distinguish, then, between what was evolved (an inherited character, by means of natural selection) and a specific response made by an individual, which adapts it, perhaps temporarily, to bright light, water, oxygen deprivation, and so on.

A major part of the process of evolutionary adaptation centers around a few primary requirements of animals and plants: feeding, metabolism, homeostasis, reproduction, escape from predators, and so on. Any large, successful group of organisms with a long history has evolved among its members a variety of different adaptations concerned with these primary requirements, and usually it is possible to recognize among them a common theme that can be traced back to the ancestor of the group, although this, of course, is often unknown. At the same time, unrelated groups of organisms that encounter similar conditions have frequently adapted to them in similar ways, producing parallel, convergent, or analogous characteristics, even though the ancestors of the groups may have been very different. Over the whole course of evolutionary history it can be seen that the various branching lines of descent show their evolutionary changes by means of a succession of adaptations for each of the major requirements of their lives (not to mention the minor ones). But the most

Figure 7. English gorse, as it grows in wet places (A) and in dry places (B). (Redrawn from E. N. Transeau, H. C. Sampson, and L. H. Tiffany, *Textbook of Botany,* 1953, courtesy of Harper & Row, Publishers.)

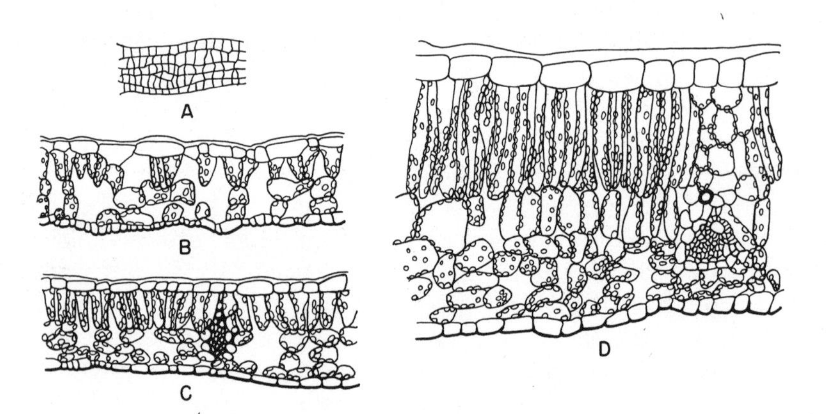

Figure 8. Cross sections of sugar maple leaves: (A) very young leaf; (B) leaf from base of a tree in shady forest; (C) leaf from middle of crown of an isolated tree; (D) leaf from side of crown exposed to sun. (Redrawn from E. N. Transeau, H. C. Sampson, and L. H. Tiffany, *Textbook of Botany,* 1953, courtesy of Harper & Row, Publishers.)

primitive members of any particular family, order, or class are already provided with characteristics not greatly unlike those of the ancestral group from which they have come. For instance, the earliest known horse, *Hyracotherium* (or *Eohippus*), differed very little from contemporary Condylarthra; the earliest Primates were scarcely distinguishable from arboreal insectivores; the earliest Amphibia were only in minor respects different from the lobefinned fishes. But in each case some new characters were beginning to appear, and when successful these would carry the new line into a long history of adaptive radiation that still further increased its specialization for new ways of living.

At first sight this may seem to imply that evolution adds new adaptations to those already present, so that descendants steadily increase in complexity of form and function compared with ancestors. This is true in a few cases, as for instance the increasing complexity of crown pattern in the molar teeth of the horse family or the complicated arrangement of muscles and joints in the limbs of tetrapods as compared with crossopterygian fins. But it is at least as common to find decreasing complexity in the course of evolution. The skulls of modern Amphibia are much simpler, have fewer bones, and fewer complications of shape than the skulls of primitive Amphibia. In several specialized groups of reptiles and amphibians the limbs and even the pectoral and pelvic girdles are reduced and lost, the bodies becoming serpentine. In many other cases organisms may evolve adaptively without any appreciable difference in complexity. Therefore new adaptations are not, in general, added to old ones. There is not an accumulation but more often replacement of old by new, or change of old characters into new ones, without adding to the total. We may speak of a succession of adaptations in an evolutionary line, but many of these are at the expense of previous adaptations that are no longer appropriate.

To illustrate these concepts more fully, and provide some evidence concerning the method of evolution, the following sections of this chapter will consider: (1) examples of the countless ways in which form, pattern, color, and behavior are adapted to deceiving predators that seek their prey by sight; (2) one example of a long-continued, highly successful adaptive radiation of wings for flying; and (3) a series of more or less complex, independent adaptations for swimming in many unrelated groups of animals.

DECEPTIVE APPEARANCE

Adaptations of visible appearance, deceptive or otherwise, occur in at least half of the known species of animals and are therefore a major accomplishment of evolution. Although this is clear enough in faunas of cold and temperate climates, it is especially impressive in the rain forests and coral reefs of the tropics, where species are most numerous and evolution is most productive.

The meaning of deceptive appearance is that animals so equipped, although visible to predators, are generally overlooked—either not noticed at all or mistaken for something that is of no interest. This kind of adaptation has long been known to field biologists and usually interpreted correctly by them, especially entomologists, for the latter are likely to have experience with a far greater number of species in their natural surroundings than anyone else, and such adaptations are common in insects.

Elements of color, pattern, form, and behavior enter almost any example, but in each case it is the adaptation as a whole that has meaning to the animal. Certain ecological factors also are involved:

1. The animal, during daylight hours, is normally in a place where it can be seen by potential predators; it is not underground or hidden in a burrow, or covered by anything.

2. The only features that take part in the adaptation are those that are in a position to be seen. If, for instance, the animal is a moth, resting with the wings outspread on a tree trunk, then only the upper surfaces of the wings, head and body take part in a general pattern of concealing coloration, but the lower surfaces, the legs, internal parts, and even that portion of the hind wings that is covered above by the fore wings, are in no way modified for concealment. Of course, the location chosen by the moth and its habit of remaining motionless by day are the behavioral elements in this example.

3. There is no necessary connection between adaptations of the adult and those of other stages of its life, which may be quite different. The eggs of a geometrid moth may be pale green, attached to the underside of a green leaf; the larva, an "inchworm" or "measuring worm," may likewise be green and almost indistinguishable from the green leaf petioles among which it lives; the pupa will have no adaptation of color because it lies in a cavity underground, where it cannot be seen.

Three of the commonest categories of deceptive appearance will serve to bring out the importance of these adaptations: resemblance to the background, resemblance to a common object, and mimicry of other, unrelated species.

Background Resemblance. The most widespread of all kinds of deceptive appearance is that relating to the background in which the animal normally occurs, making the animal inconspicuous or even impossible to recognize unless it moves. Common backgrounds on which cryptically colored animals live include the bark of trees or logs, dead leaves on the ground, dead or green leaves on trees or bushes, grass, shrubbery, the corollas of certain flowers, lichens growing on trees or rocks; sand, gravel, rock, either in or out of water; corals and other marine growth in the sea; water as seen from above or from within; the sky as seen from under water; snow.

Across any of these backgrounds there may travel numerous animals that are clearly visible or that, by moving, make themselves apparent. Crows are readily seen in almost any situation. Zebras in their normal habitat are conspicuous for a long distance (in spite of some attempts to explain their stripes as deceptive). Many kinds of butterflies, beetles, birds, fishes, and so on, depend on no adaptive coloration whatever; their survival and reproductive success are due to other kinds of adaptation: alertness, quick response, speed and erratic flight, armor, nauseating odor, sting, bite, or other defense. Nevertheless, a careful search of a few tree trunks in a forest will reveal cryptically colored moths, beetles, spiders, bugs, and in some places tree frogs and lizards that are invisible a short distance away. Usually they remain motionless until actually touched. Walking among the fallen brown leaves of a tropical forest one is continually starting up brown grasshoppers, brown moths, brown butterflies, mantids, beetles, frogs, toads, lizards, spiders, and so on, none of them resembling a dead leaf, but all blending, while motionless, into a dead-leaf background in such a way as to escape notice. Details of their patterns include leaflike parts, such as a midrib and veins coming from it, holes or notches like those that are made by insects eating a leaf, blotches resembling fungus spots, and so on, which have to be seen in the field to be appreciated. A very common detail is a pale brown area sharply edged, with a dark brown or black portion adjacent, as would be seen where one leaf lies partially over another and shadows it.

The number of kinds of green insects that are concealed among leaves of trees, bushes, and shrubbery can best be realized by making a collection around electric lights at night in a forest, especially in the humid tropics, for it includes green moths, green beetles of various families (some of which do not come to light, however), great numbers of green Hemiptera and Homoptera, Orthoptera, especially the leaf mantids and long-horned grasshoppers, and even green cockroaches. Naturally the green caterpillars will not be obtained this way, nor will the green tree frogs, lizards, and snakes. Again, it is not resemblance to a single leaf but to a copious environment of green that results in making these animals difficult to find by day.

Regarding the other backgrounds listed above, each has its quota of animals concealed by their resemblance to it, and the number of individuals and of species in any ecological association of such animals is much larger than would be possible if they were all readily visible to predators. In the surface waters or shallow littoral zone of the sea the same principles apply to many groups of fishes, crustaceans, and other organisms, as long as there is ample light for them to be visible by day. In the abyssal depths coloration presumably has no deceptive function; organisms there are commonly black, red, gray, or white, and visibility is limited to objects that are luminous.

In the past, doubt has been expressed by some biologists of the basis, or even the reality, of these adaptations for concealment from predators. For example, in the years before the genetic basis of natural selection was understood, it was sometimes supposed that animals could in some way take on the color of their background by means of visual impulses sent from the eyes to the central nervous system. Indeed, there are a few groups (chameleons, some tree frogs, flounders, and certain squids and octopi) that change color and pattern according to just such stimuli and are able to match various common backgrounds on which they may occur. Such, however, are highly specialized physiological adaptations serving the cryptic function, and based upon movement of pigments within various chromatophores (color cells) of the skin, under control of motor nerves, whereas the colors and patterns of the animals previously considered are determined during development by a genetic mechanism and cannot be changed by putting the animal in a different situation.

Another kind of objection sometimes expressed is that our judgment of the effectiveness of this deception is based on human experience and may not correspond with that of birds, monkeys, or lizards. Are they actually deceived in the same way? Do they not eat those with a cryptic pattern as well as any others? Have we actually seen such predators feeding on prey that is conspicuous to us but overlooking those that we find difficult to see? Do dark- and light-colored moths stay, respectively, on dark and light tree trunks? In the case of variable species, how can we decide just what degree of selection is carried out by the predators? What is the evidence that selection produces more perfect adaptation? How many ornithologists have watched birds selecting their prey on tree trunks?

Although, as described by Cott (1940) in his extremely comprehensive study of adaptive coloration, many experiments had already been performed at that time, the whole problem has now, since 1940, been investigated in great detail, especially in England, by E. B. Ford, H. B. D. Kettlewell, and others. The results, covering both the selective and genetic aspects, are reviewed by Ford (1964) in his book *Ecological Genetics.* Briefly, the subject is a recent, rapid evolutionary change in the coloration of moths inhabiting tree trunks in England. The species belong mostly to the families Geometridae and Noctuidae (in the broad sense). The adaptation is to a new ecological situation produced by man—the darkening of tree trunks and extermination of lichens that normally grow on them by soot from factories in the industrial cities, most of which are in the Midlands. The smoke does not scatter at random but is carried toward the east and northeast by the prevailing westerly wind, thus affecting nearly all the area to the shore of the North Sea; no darkening of the trees has occurred in western England except in a few limited localities.

In these species of moths, which normally have light gray or whitish wings with dark lines and spots, mutations occur that make the wings and body black or dark gray with little or no pattern. Such mutations, producing melanic individuals, are not particularly rare among animals in general, and they happen from time to time in many kinds of insects. In this case, as illustrated by *Biston betularia,* the peppered moth, melanism was first noticed in the middle of the nineteenth century (1848) in Manchester, gradually increased in frequency, then more rapidly; by 1895 it was the normal color of the species (about 98 percent) in the Manchester area and has now become abundant in other industrial parts. But it is not found in western England or in western and northern Scotland, which are free of deposits of factory soot (Fig. 9).

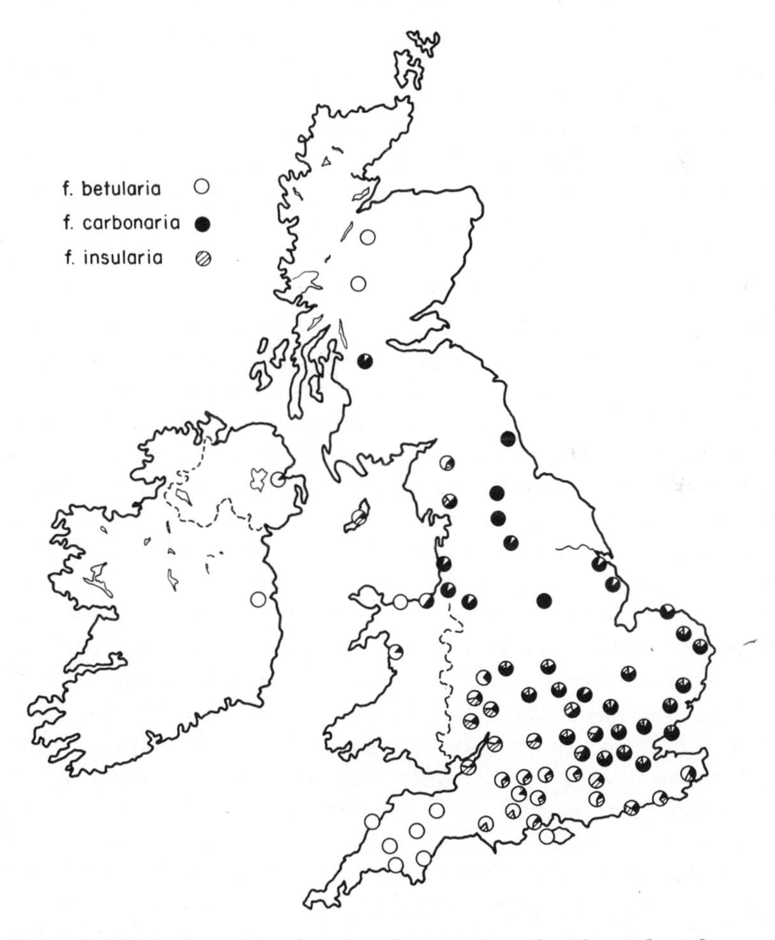

Figure 9. Map of Britain showing frequencies of pale and melanic forms of *Biston betularia,* according to work of Kettlewell; the dark forms are in industrial areas and east of them but not west, because of the direction of the wind which carries the pollution.

The melanic form is *carbonaria;* another dark form, *insularia,* also appeared as a mutant somewhat later and is now common in parts of England. (Note that the soot is in no way responsible for any mutation; the darkened tree trunks merely provide a favorable situation for survival of dark moths.) It has been shown by breeding experiments that both mutants are dominant to the pale form and, especially *carbonaria,* are hardier, more viable than it is. Kettlewell carried on numerous experiments in forests where the trees were darkened by soot and others where they were not. Equal numbers of moths of the pale and dark types were released on the tree trunks and were watched with binoculars until all of either the light or the dark had been removed. Birds feeding on them in the sooty area were the English hedge sparrow, robin, and redstart, which examined the tree trunks and caught the moths. On the dark trunks the birds found many more of the conspicuous pale moths. Redstarts, for example, ate 43 *betularia* to 15 *carbonaria.* In the opposite situation, where equal numbers were released on clean, light-colored, lichen-covered trunks in Dorset, five species of birds captured 164 *carbonaria* and only 26 of the pale *betularia.* Moving pictures were taken of the birds in action, giving conclusive evidence for a common kind of behavior that apparently had not been witnessed before by ornithologists or entomologists.

Still other kinds of experiment were devised. Several hundred male moths were released and later recaptured by their response to the chemical attraction produced by females and also by trap lights. The results showed that the cryptically colored moths returned in much larger proportions than those that were conspicuous on the tree trunks. With this and much other information obtained in recent years it seems clear beyond doubt that a strong selective pressure is maintained by predators to eliminate any but the most cryptically colored individuals. In fact, a few of the cryptic are taken and a few of the conspicuous are overlooked, but the pressure favoring one instead of the other is very strong. Kettlewell also showed that moths alighting on a tree trunk frequently move to adjust their positions to match most nearly their own colors before settling down, although they do not go so far as to fly to a differently colored tree.

This study has shown that rare mutations, occurring at random, may become fixed in a population by natural selection under favorable conditions, and thereafter cause the species to be polymorphic (two or more genetically and phenotypically different forms in the same population). "Industrial melanism" has been called *transient polymorphism* because, so far at least, it has not settled down to a constant ratio of the different forms to each other. It is a thoroughly documented case of evolution in action, at a rate far more rapid than was previously thought possible.

Object Resemblance. A second category of deceptive appearance is that in which the animal resembles an object, common in its environment, so closely as to be mistaken for it. Animals with this adaptation are not concealed or overlooked; their success depends upon being noticed as an object but one that is of no interest to the predator. Not only color, pattern, and behavior, but in this case form, including appropriate size, are essential to the deception. The object imitated is a common, familiar part of the environment. For instance, membracid bugs (tree hoppers) that resemble thorns are found among such thorns, not on the ground, not on trees that lack thorns, and not even on trees with other types of thorns.

Many insect collectors have had experiences such as the following. On a trail in the dense forest of Panama, at dusk, I was looking for insects. A fairly large pale brown insect flew across the trail and I thought I succeeded in catching it in the net, but on reaching into the net I found only a dead twig that had apparently fallen in by mistake. On picking this up to throw it out, I found that the "twig" had legs and wings and was struggling to get away. It was, in fact, a phasmid, or "walking stick." Among other stick insects obtained in Panama was one, some 6 inches long, in which the body was like a rough, blotchy, dark-gray broken stick; at the hind end a couple of fringes of thin bark hung where the twig had apparently broken from the tree; there were two or three short thorns, and a number of irregular pale spots suggesting the growth of mold. The front legs, held in line with the rest of the body anteriorly and pressed against each other, had at the bases inwardly, shallow excavations where the bulging eyes fitted; thus no space showed between the legs in front of the head, and the apparent twig was about 9 inches long (body plus forelegs). The middle and hind legs were, of course, branches of the main twig. Like most stick insects, when placed on a branch it hung, not rigidly, but with a gentle, rhythmic swaying motion like that of a real twig swinging in the breeze. Such astonishing adaptations as this seem to justify the remark commonly made that they are "too good to be true," they are "unnecessarily perfect," or "how could natural selection go so far beyond any adaptive advantage?" They do exist, however, and abundantly. Our judgment concerning natural selection must be qualified by understanding the actual selective work of the predators concerned. This is discussed, with other questions, in the following section.

Among many other objects that serve as models for detailed resemblance, leaves, usually brown, are imitated by several genera of butterflies, the most famous being *Kallima* of the Old World tropics (Fig. 10) and its close relative *Doleschallia;* also by the South American freshwater fish *Monocirrhus,* living in clear water where leaves fall in and drift slowly along, and by various leaf insects: *Phyllium, Chitoniscus,* katydids, *Cycloptera,* and so on. Certain caterpillars imitate twigs, petioles, grass stems, and pine needles. It is common to find a caterpillar, beetle, or spider that resembles in detail a fresh bird dropping, and there are small greenish-brown Chrysomelid beetles that imitate caterpillar droppings.

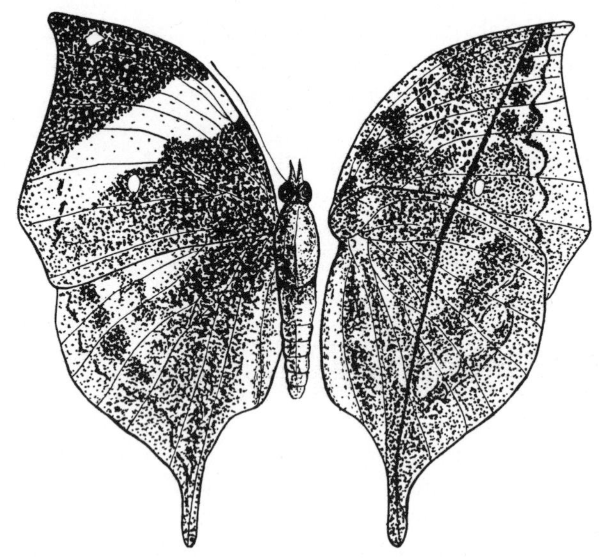

Figure 10. Leaf butterfly of India, *Kallima inachus,* upper and lower surfaces.

Mimicry. Among the numerous common objects in the environment of almost any animal are other animals, many of which do not differ greatly from it in size and general appearance. If some of these, although easily seen, were for any reason avoided or disregarded by predators, we might expect them to serve as models for imitation by various species, in the same way that thorns and twigs, leaves, and bird droppings are imitated. In fact, this is very commonly seen. It is the well-known phenomenon of mimicry.

Evidently, just as with the unattractive or distasteful objects already mentioned, the resemblance of an edible species to one that is inedible gives it some relief from predation, if the resemblance is sufficiently close to deceive a predator. And, just as the imitation of a dead leaf or a stick may extend to extraordinarily fine details, provided they are visible, so mimicry of an unappetizing model by unrelated species may extend to precise details of color, pattern, shape, and actions (the latter including a buzzing noise in the case of some mimics of Hymenoptera). Again, in this situation one of the essential elements is that the model be present, and common, in the area occupied by its mimics; usually it is much more common than they. Another is that the mimicry does not include features that are not visible (or audible) to a predator, or parts of the life history in which other, quite different, adaptations may prevail. An additional element is that predators actually become aware of the repellent qualities of the common model.

Here some examples are necessary. Insects that sting are all female members of certain families of Hymenoptera; the sting is an ovipositor (egg-laying organ) to which a poison gland has been added. Stinging Hymenoptera are almost always conspicuous in their color and behavior. Their actions may include loud buzzing, vibrations of antennae, flicking of the wings, or quick, jerky motions. There are many thousands of species, most of them tropical. They provide an abundance of models for mimicry by flies, beetles, certain day-flying moths, and other nonstinging insects, each less numerous than its model. For every distinctive hymenopterous pattern, or "image," such as large bumblebee, small bumblebee, honeybee, mudwasp, yellow hornet, black and white hornet, velvet ant, red ant, and so on, there are several unrelated mimics of it in every geographical region where it occurs.

For the mimics to benefit from the deception it is necessary that birds, monkeys, lizards, frogs, toads, and so on, be made aware that the object "bumblebee" means a painful experience if they try to eat it, and also that this association be remembered, even if the same image is presented, later, by a fly or a moth. Some biologists have questioned whether the predators mentioned could learn to recognize the models by a few experiences and then remember them, and whether they were, in fact, deceived by mimicry. Among many experiments to test this is a series performed by J. V. Z. and L. P. Brower (1965) using toads (*Bufo*) as the predators, honeybees (*Apis mellifera*) as the models or disagreeable objects, and droneflies (*Eristalis*), which mimic honeybees, as the harmless edible species. Without previous experience toads took the latter freely. Toads fed on honeybees quickly refused them, after one or a few trials, and thereafter, for several days, remembered and continued to refuse them, after having been fed other food in the meantime. (A small minority of the toads apparently were unaffected by stings and accepted the bees anyway.) Droneflies given to experienced toads were then in most cases refused, to an extent that showed highly significant effectiveness of the mimicry.

Similar experimental work by J. V. Z. Brower (1958a, b, c) on the effectiveness of mimicry in certain butterflies of North America, in which the Florida scrub jay (*Cyanocitta c. coerulescens*) was the predator, produced essentially the same results. The experiments were intended to test the suppositions that (1) the monarch, *Danaus plexippus,* is a distasteful model mimicked by the viceroy, *Limenitis a. archippus;* (2) the pipe-vine swallowtail, *Battus philenor,* is a model for mimicry by *Papilio troilus* and by the females of *P. glaucus* and *P. polyxenes;* (3) the queen, *Danaus gilippus berenice,* is likewise mimicked by the Florida viceroy, *L. a. floridensis.* In all these, both the upper and lower surfaces of the wings enter into the supposed mimicry. *P. glaucus* is the yellow-and-black-striped tiger swallowtail, both sexes having the tiger pattern in the north, but in the southern states females are dark blue, with a pattern like *B. philenor;* similarly, it is only the female of *P. polyxenes* that resembles *philenor,* the male being black with yellow markings. The supposed model in each case is a conspicuous, strongly colored insect, not a particularly agile flyer, and has a distinct taste that is presumed to be offensive to a predator.

The experimental birds ate, without hesitation, palatable, nonmimetic butterflies of other kinds that were presented throughout the experiments. The models were at first eaten, or pecked, and this continued irregularly but were then usually (not invariably) rejected on sight alone. After "learning" the monarch, no bird ate a viceroy, although control birds (not having tried the monarch) did so. All the mimics of *philenor* were shown to be edible and acceptable to a series of control birds. After experience with *philenor,* which was rejected even more than the monarch, *P. troilus* was almost always rejected on sight, and in most cases the female of *polyxenes* also. But the female of *glaucus* caused the birds some confusion; it was accepted and rejected about equally. It is, in the south, a larger butterfly than *philenor,* and the resemblance is not quite as close as in the other species. Apparently the birds were taking it as different in some cases.

In the third experiment some of the previously experimental birds were used as controls and refused the Florida viceroy, even though it was shown by another bird to be acceptable and edible. *L. a. floridensis* is quite easily distinguishable from *L. a. archippus* and from the monarch, but the general resemblance is sufficient to make most of the birds unwilling to accept it. This indicates that after thoroughly learning a model, birds will "generalize" and refuse anything that is even vaguely like it. Thus mimicry can have adaptive value even when the resemblance is not yet precise, provided the mimics are much less common than the models.

The common type of mimicry just described (Figs. 11 and 12), in which a repellent model is mimicked by various edible species, is called *Batesian mimicry,* after Henry Walter Bates (see Bates, 1862), who recognized many examples of it and correctly explained it during his years as a naturalist on the Amazon. Later (Müller, 1879) other cases of mimicry were found in which many of the species concerned were characterized by some disagreeable features. These are now known to constitute a fairly important category, called *Müllerian mimicry.* Apparently convergence toward a common pattern occurs among all members of a Müllerian group, whereas in Batesian mimicry the mimics differ from their own nearest relatives, but the model is unaffected. One interesting consequence of mimicry in general is that it allows many more species to exist in a given locality than would appear to be there, as far as the discriminative powers of a predator are concerned. The significance of Müllerian mimicry is that it reduces the number of different "images" that a predator needs to learn to avoid attacking those species that give it an unpleasant experience. By the predator's early recognition of the common pattern a selective advantage is given to increasing accuracy of the resemblance among members of the mimetic group. It has been found, as we might expect, that large mimetic groups among tropical butterflies may include both Müllerian and Batesian types.

Figure 11. Above, *Heliconius erato,* a distasteful South American butterfly, mimicked by a day-flying moth, *Pericopis phyleis* (second from top). Third, *Methona confusa,* another South American butterfly, mimicked by *Castnia linus* (bottom), which unlike most other moths, has swollen tips to its antennae. (Redrawn from R. A. Fisher, *The Genetical Theory of Natural Selection,* Dover Publications, Inc., New York, 2nd ed., 1958, with the permission of the publisher.)

INSECT WINGS

Among invertebrate animals the adaptation for flight, as self-propelled locomotion through the air, probably originated only once. Various small or microscopic animals are transported by the wind on occasion, and frequently this is a normal part of their lives. Cysts of

Figure 12. Above, *Danaus tytia,* a widespread distasteful Oriental butterfly, mimicked by *Papilio agestor* (middle) and *Neptis imitans* (below), in different families. (Redrawn from R. A. Fisher, *The Genetical Theory of Natural Selection,* Dover Publications, Inc., New York, 2nd ed., 1958, with the permission of the publisher.)

Protozoa, rotifers, various soil inhabitants, and so on, are commonly blown about. Newly hatched spiders climb to the tip of a twig or blade of grass, release a silk thread, and when it is long enough, are borne away by it as if by a kite. Some pelagic squids leap from the ocean. But only the winged insects, among invertebrates, can fly.

There is a certain amount of evidence on the origin and early evolution of insect flight. Several orders of flying insects are known from rocks of the Pennsylvanian period and later. The structure of their wings is much as would be expected from the study of morphological relationships among living insects. We can follow, in fossil and living orders, the diversification of a fairly complex character into numerous radiating lines of specialization which occasionally become parallel or convergent among themselves. There can be no doubt that the ability to disperse by flying and to carry on many activities in the air has contributed to the great predominance of insects over all other classes of animals in numbers and diversity.

There are several groups that have no wings at any stage, and these fall into two categories:

1. *Primitively wingless* (Fig. 13A). The Protura, Collembola, and Thysanura are three orders of small, flightless insects. They survive today because of their adaptation to concealed, moist habitats on the ground and in the humus layer of the soil, where they feed on dead organic matter.

2. *Secondarily wingless* (Fig. 13D). The Siphonaptera (fleas), Anoplura, Mallophaga (sucking and biting lice, respectively), and numerous families or smaller groups among several other orders are wingless but carry in their structure or life history evidence of descent from winged ancestors. Thus it is probable that the fleas have been derived from two-winged flies (Diptera); the larva and pupa as well as the general structure of the body of a flea are essentially those of Diptera. There are wingless beetles in many families, wingless bugs (Hemiptera), crickets, wasps, moths, and so on. In most of these cases some intermediate kinds can be found in which the wings are not completely lost but are reduced.

When the flying orders are compared with each other, they show three general conditions: (1) two pairs of wings that are essentially similar although they may differ in details, the first pair being on the mesothorax (second segment of the body) and the second pair on the metathorax (third segment); (2) only the first pair is present, but the second may be represented by a vestigial structure, such as the halteres of a fly; and (3) only the hind wings are present or functional, whereas the front pair may be vestigial or represented by protective wing covers, such as the tegmina of a locust and the elytra of beetles (Fig. 14). In several orders the fore and hind wings are virtually identical in shape, function, and the detailed pattern of veins, such as dragonflies (Odonata), termites (Isoptera), and Neuroptera. But in others, such as Homoptera, Lepidoptera (moths and butterflies), and Hymenoptera (ants, bees, and wasps), there is a pro-

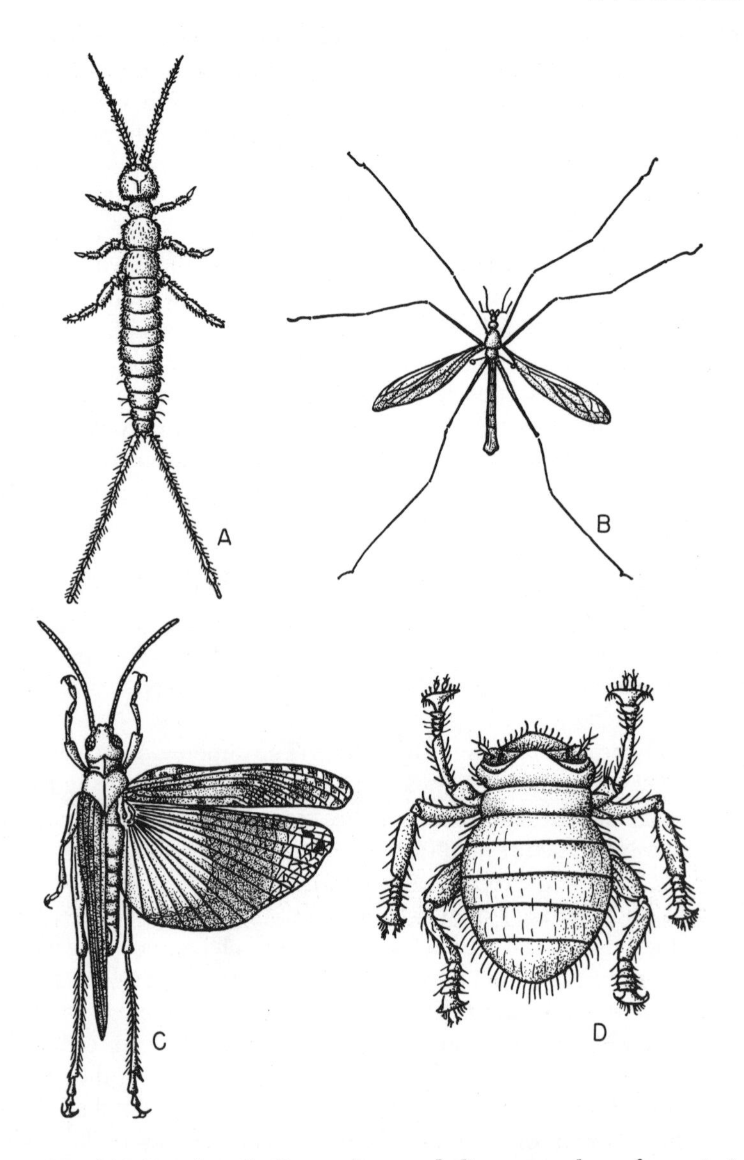

Figure 13. (A) Bristle-tail, *Campodea staphylinus,* member of a primitively wingless order of insects, Thysanura. (B) Crane fly, *Tipula,* one of the Diptera, in which the hind wings are represented only by short, club-shaped halteres. (C) Locust, *Spharagemon bolli,* with two pairs of wings, the first pair forming covers for the hind wings when at rest. (D) Bee louse, *Braula caeca,* a wingless, parasitic fly (Diptera) found on bees. (Redrawn from J. H. Comstock, *An Introduction to Entomology.* Copyright 1920 by Comstock Publishing Company. Copyright 1924 by J. H. Comstock. Copyright 1933, 1936, 1940 by Comstock Publishing Company, Inc. Used by permission of Cornell University Press.)

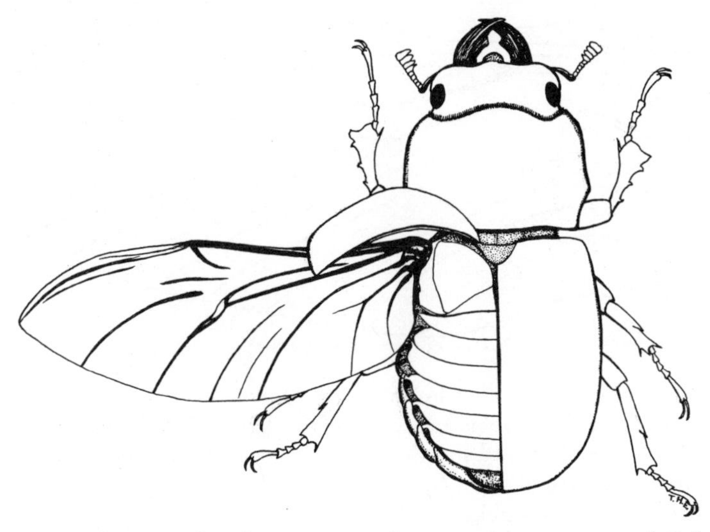

Figure 14. A stag beetle, *Lucanus*, showing characteristics of Coleoptera.

gressive increase in differentiation between the two pairs of wings, although in some members of each group the wings are still much alike. Thus it seems clear that among existing winged insects the most primitive condition is shown as two pairs of closely similar wings.

This is borne out by other characters. In Lepidoptera, for example, the most primitive condition of the wings appears in three families that are placed in a suborder Jugatae; the first of the families, Micropterygidae, contains some minute moths in which, unlike any other Lepidoptera, there are still functional mandibles, used in feeding on pollen. The wing pattern in Jugatae is much like that of Trichoptera (caddis flies), from which Lepidoptera were derived; the larva of a caddis fly is remarkably similar to a caterpillar, and the pupa could readily be ancestral to that of Lepidoptera.

The primitive pattern of wing venation in insects was worked out many years ago by Comstock, Needham, and others (Comstock, 1918). In individual development this pattern (Fig. 15) is based on arrangement of the branching air tubes or tracheae within the saclike wing buds that develop just before the last ecdysis, that is, in the pupa stage in most insects; but it is also like the pattern in the immature, functionless stages of wing development in the early instars of Orthoptera (grasshoppers, locusts, and their relatives). Further, it is basically the pattern found in cockroaches, and these are among the earliest of all fossil insects, yet are still common today; they appeared in the Pennsylvanian (late Carboniferous) period, some 250 million years ago. In the Pennsylvanian and

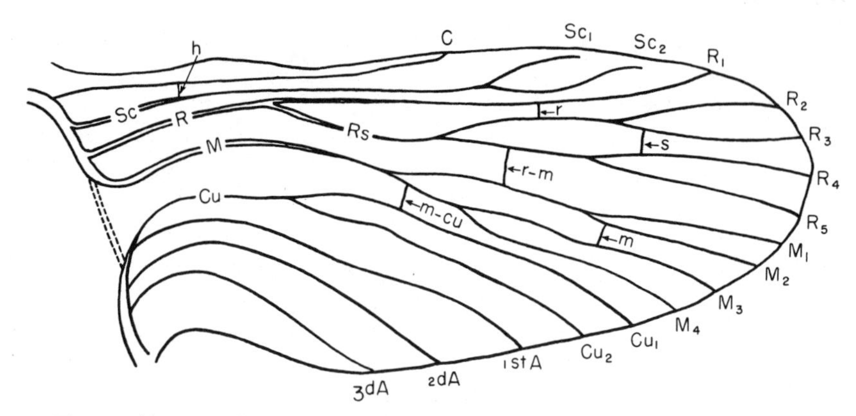

Figure 15. Hypothetical primitive type of wing venation of insects, with the named crossveins added. (Redrawn from J. H. Comstock, *An Introduction to Entomology*. Copyright 1920 by Comstock Publishing Company. Copyright 1924 by J. H. Comstock. Copyright 1933, 1936, 1940 by Comstock Publishing Company, Inc. Used by permission of Cornell University Press.)

Permian came several orders now extinct, some of which were probably ancestral to the more advanced modern orders. In one strange group, Megasecoptera, the prothorax or first segment carried a pair of winglike flaps, as if these insects had experimented with three pairs of wings but finally settled for two (Fig. 16). For a different view, see Edmunds and Traver (1954).

Several inferences may be drawn from this brief summary, although the story would remain incomplete without more detailed study of wing venation in each order, and supporting evidence from the evolution of mouthparts, body and head structure, and metamorphosis. Effective flight with two pairs of wings must have originated in the Pennsylvanian or earlier, and it proved to be an enormous advantage in dispersal and probably in mating of adult insects. The young after hatching began to produce wingbuds of increasing size during their successive stages of growth. At first no actual metamorphosis took place. The young grew by steps, each increase accompanied by a shedding (ecdysis) of the chitinous exoskeleton, the last stage being the sexually mature adult, able to fly. Such differentiation of adult from immature stages could lead the way to differentiation of habits and behavior, and therefore to taking advantage of more than one ecological opportunity during an individual lifetime. If the ecological difference were slight, as in grasshoppers, then wings might continue to appear in stages of increasing size up to the adult; if it were great, as in the emergence of mayflies and dragonflies from nymphs that live in the water, such rudimentary wings would remain concealed and not appear until the final transformation. But there is no pupa in these insects or in the Hemiptera, Homoptera, or various other orders that go through a partial metamorphosis.

We see the beginning of complete metamorphosis in such insects as the dobsonflies (Megaloptera), in which a rather wormlike larva without wingbuds forms a cocoon or cell, in which it enters a quiescent stage, shedding its last larval skin and becoming a semipupa; the legs, mouthparts, antennae, and wings are easily seen but the latter have not been fully developed. Finally, it may have happened in more than one line that the actively feeding, rapidly growing larva transformed to a quiet pupa in which most of the internal structure was broken down, then literally remade into the quite different limbs, mouthparts, sense organs, reproductive system, and wings of the prospective adult, this emerging in a final ecdysis from the pupa. Trichoptera and Lepidoptera in one instance, Diptera and Siphonaptera in another, Coleoptera, and finally Hymenoptera are all characterized by complete metamorphosis, probably arising in four separate stocks.

Specializations, such as the absence of hind wings in Diptera, or of any wings in Siphonaptera, modifications of fore wings to wing covers in Coleoptera or hemelytra in Hemiptera, all seem to have taken place in groups that now constitute independent orders, not directly related; therefore these changes were evidently based on favorable variations that occurred occasionally in ancestral four-winged insects. Their adaptive value that may have encouraged survival is often impossible to determine, but many examples of such variations occur in limited groups and show obvious adaptive meaning. One such is the frequency of wingless beetles on the smaller oceanic islands, where flight is an obvious hazard because of the persistent wind; normal individuals of many families that fly are

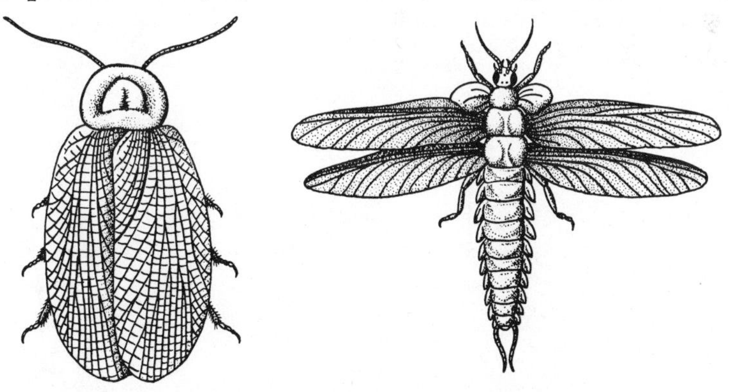

Figure 16. Primitive insects of the Pennsylvanian period; a cockroach on the left (much like recent cockroaches), and on the right an insect (*Stenodictya*) with "wings" on prothorax as well as the usual mesothorax and metathorax. (Redrawn from C. O. Dunbar, *Historical Geology*, 1949, courtesy of John Wiley & Sons, Inc.)

more likely to be lost by blowing out to sea. At one point of the coast of California a small local colony of *Drosophila* shows a remarkably high proportion of individuals with vestigial wings or no wings, apparently for the same reason.

SWIMMING ADAPTATIONS

It may be best to recognize here several fundamentally different mechanisms for swimming, showing (1) that they have no direct relationship to each other and originated in quite different groups of animals, but that (2) in discussion of each kind of mechanism, it is exemplified by animals of various groups which may or may not be directly related; if not, they are convergent with respect to the adaptation in question.

Ciliary Swimmers. Among animals that swim by means of cilia (Fig. 17), an important group is the ciliate Protozoa; in the order

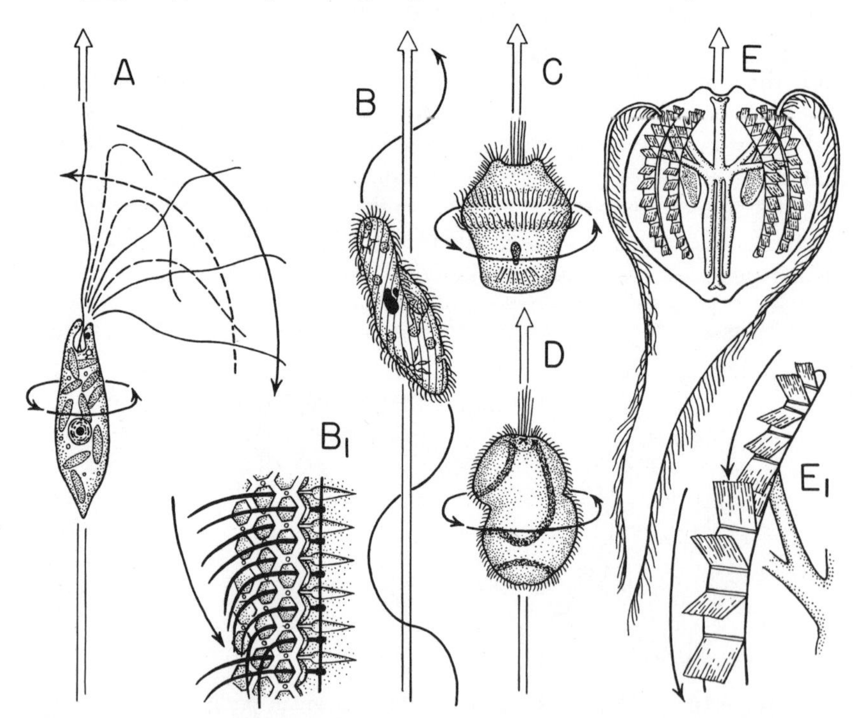

Figure 17. Flagellated and ciliated swimmers: (A) *Euglena;* (B) *Paramecium;* (C) trochophore larva of a mollusk; (D) tornaria larva of *Balanoglossus,* a hemichordate; (E) *Pleurobrachia,* a comb jelly, Ctenophora. Part (A) redrawn in part from C. A. Villee, W. F. Walker, Jr., and F. E. Smith, *General Zoology,* 1958, courtesy of W. B. Saunders, Inc.

Holotricha, the cilia cover the whole surface and run in lines diagonal to the axis of the body. Most of the members, such as *Paramecium,* swim in a spiral course with the oral end forward, twirling around and around as if they were going over and under an imaginary straight line. The beating of the cilia is very rapid. Some Holotricha have distinct rows, others do not show them clearly. Among the larger kinds, such as *Opalina,* one can see the beating of cilia in waves. In *Paramecium* and others, there is a system of minute threadlike connections underneath the surface of the body between the base of one cilium and the next. These threads apparently act as a coordinating system.

A planula, the microscopic larva of a jellyfish, is similar to a ciliate protozoan, except that it is multicellular and has no mouth.

Trochophores, ciliated larvae of many of the marine invertebrates, are transparent and swim in the open ocean. A trochophore is shaped like a top and often maintains a position in which the apical tuft is up. Around the equator of the body are one or several bands of cilia that vibrate continually, as do those of the apical tuft. The result is a twirling motion, in which the animal moves in the general direction of the apical tuft, but it may, like a protozoan, travel in a spiral.

A tornaria is also a microscopic transparent swimming larva; it transforms into an acorn worm, one of the hemichordates, such as *Balanoglossus.* It is essentially like a trochophore, except that the ciliated bands around the body are not in a circular position but are curved or bent in various directions. Usually there are two or three such bands, and the body, although superficially round, like a top, can easily be seen to have bilateral symmetry; the mouth is on the ventral side, halfway down.

The ctenophores, or "comb jellies," are bilaterally symmetrical in some details but superficially usually appear to be radial. Most of them have bodies composed of transparent jelly like that of a medusa, but they differ from the jellyfishes in their locomotion. They swim by a slow, even, gliding motion, not by contracting and relaxing. The movement is produced by eight longitudinal bands of cilia that can easily be seen by the naked eye, running down the sides of the body, from near the apical end. The cilia in these bands can also be seen; in reflected light their wavelike motion is easy to make out, and in many cases the cilia of each band are partially or completely fused in transverse rows. The beat is such that it seems to travel down the band in waves. Most kinds of ctenophores have a pair of tentacles that are let out of openings, one on each side of the body, and they may dangle to several times the length of the body. When touched, they are instantly jerked up into the holes, and then can be seen to relax and emerge again. These are not used in locomotion. The movement of the ctenophore is forward in a line toward the apex of the body.

Very few general statements can be made about animals that swim by means of cilia, but one is that nearly all of them are of microscopic size, and this means that the effective viscosity of the water through which the animal moves is relatively great. During swimming the cilia move in continuous rapid beating, whipping violently down from the vertical position in one direction, but returning somewhat more slowly. This results in a thrust that is stronger in the direction of the first push. Since the cilia are minute in relation to the size of the animal, there must be a great number of them, and to maintain a continuous motion of the body against the water, the beating of the cilia has to be coordinated by an internal apparatus. The only ciliated swimming animals not of microscopic size are the ctenophores. Any other animal as large as a ctenophore and able to swim would be expected to develop a muscular system for movements of the body, but apparently the ancestors of ctenophores were ciliated animals of small size, and adaptation of the cilia as bands serving a locomotor function was successful.

Jet Swimmers (Fig. 18). Several different unrelated groups of swimming animals have a mechanism that can be called a pump. Ordinarily living pumps are rhythmic in action, for after every contraction that ejects water, there has to be relaxation to fill the chamber. In the case of a medusa or jellyfish, the normal position of the body is with the convex surface of the bell upward and the margin hanging down, with tentacles dangling from the edges, but a jellyfish has very little control over its position other than that provided by the dragging of the tentacles in the water. The contraction of the jelly is by a circular constriction of the bell, making the downward-facing cavity smaller and the upper surface more convex. In small jellyfishes this movement is rapid and energetic, but in many of the large kinds it may be slow and feeble. As a result, the smallest jellyfishes appear to move around in the water in a much more

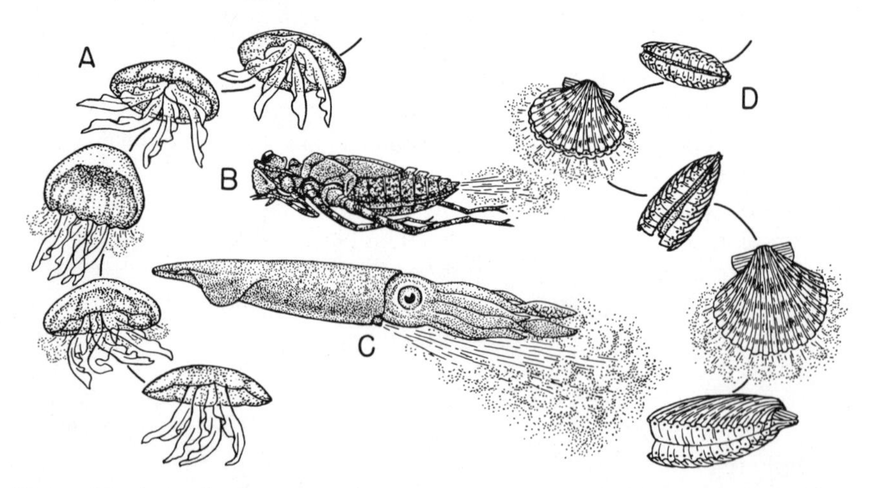

Figure 18. Animals that swim by pumping or jet propulsion: (A) medusa (jellyfish); (B) nymph of a dragonfly; (C) squid; (D) scallop (*Pecten*).

lively fashion than the large kinds, which hardly do more than maintain their position or slowly rise. Because of local disturbances in the water or irregularity in the beat, jellyfishes usually do not move straight up but tip to one side or another and propel themselves with each beat in a slanting upward direction.

Among mollusks, there are two groups in which a pumping mechanism for locomotion has developed. One includes the scallops, genus *Pecten*. The animal, shaped like a clam but more symmetrical, rests on one of its two shells with the hinges lying horizontally and the upper shell slightly separated from the lower. This provides a wide opening for water to enter the mantle cavity, and around the edges of the mantle are sensory fringes and a row of eyes. When disturbed by touch, vibration, or a shadow, the scallop reacts by suddenly closing its valves by adductor muscles; this is repeated rapidly several times, each time shooting a jet of water out of the opening. As a result, the scallop darts erratically through the water until, tired of its action, it comes to rest at another place on the bottom. It has little or no control over the direction in which it goes, but the quick flapping of the shells against each other carries it out of reach of the original stimulus.

The other group of mollusks that use a pump in swimming is the Cephalopoda: the octopus, squid, and nautiloids. These animals primitively had a chambered shell usually in the form of a spiral, the outermost chamber being occupied by the body of the animal, and the mantle cavity opening on the ventral side by a siphon or tube that could be turned in any direction, forward or back. The nautiloids have a chambered shell, but it has been reduced in the squids to an internal bladelike structure and is lost in the octopus. By contraction of the muscular mantle chamber and turning the siphon in the direction of the tentacles, the animal emits a strong jet of water that propels it backward in the opposite direction. An octopus is not well streamlined and uses this mechanism for escape in emergencies, but a squid, on the other hand, is beautifully streamlined and spends most of its life in active swimming, forward and back. A horizontal tail fin at the posterior end undulates and can be used as a rudder. The body comes to a point at that end, but it is also pointed at the anterior end when the tentacles are placed closely against each other in front of the mouth. Then by moving the siphon so that it aims backward, and ejecting water, the squid moves forward, sometimes with great speed. Many squids that swim in the open ocean leap out of the water at night and land on the decks of boats, if there is a light that attracts them. Therefore they must have carried the pumping mechanism to a high perfection, for it enables them to escape predators in a manner analogous to a flying fish.

Jet propulsion occurs even in insects. A dragonfly nymph crawls around on the bottom or on stones and vegetation in fresh water. It has a pump at the posterior end which functions both for breathing and for locomotion in emergencies. This is an enlarged rectum in which the inner surface is lined with folds that serve as gills for breathing. The muscles of the posterior part of the abdomen have become especially developed to act as a pump ejecting water violently, so that the insect will dart away if disturbed, without using its legs, which are held close against the body. Most dragonfly nymphs are not well streamlined, but this mechanism serves for short and sudden escapes. The action is usually not continued, but after one or a few squirts the insect comes to rest elsewhere on the bottom.

Leg Swimmers (Fig. 19). A great many animals make use of paired appendages for swimming, and usually the adaptation is a specialization of the walking mechanism, but an exception to this is seen in the polychaete worms, such as *Nereis*, which did not have walking ancestors. The paired appendages are nonjointed flaps of rather complex form, the parapodia. There are two bristles on each, an upper and a lower, which would be useful in burrowing in the mud. There is a partially separated upper lobe that has the function of a gill, being supplied with blood vessels. Internal stiffening threads run from each parapodium into the body, and these help sustain the organ in its projecting position, but evidently it is simply an extension of the body wall rather than an independently movable limb. While swimming, the undulations of the body are transmitted to the parapodia and intensify the pressure of the body against the water so that the *Nereis* swims with speed and agility. The undulations seem to be in any plane and not just from side to side, although that is the strongest emphasis. The position and movement of the appendages are not oriented toward walking on the substratum but

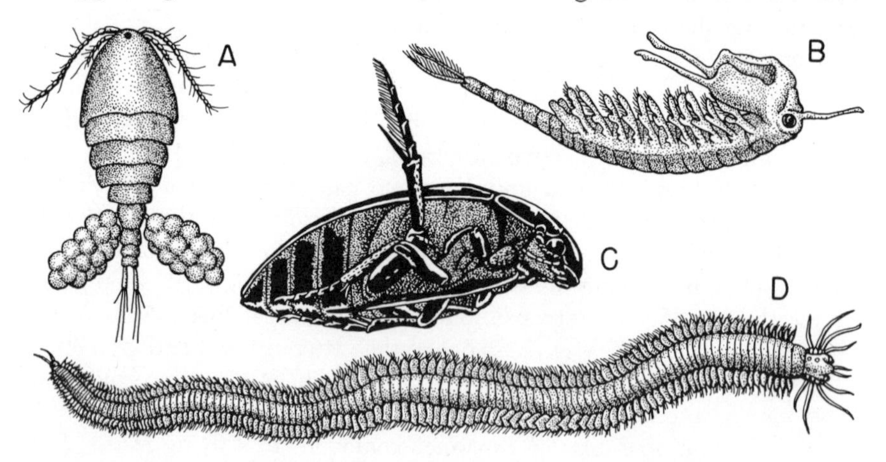

Figure 19. Animals that swim by kicking or jerking their appendages: (A) copepod, *Cyclops;* (B) fairy shrimp; (C) water beetle, *Hydrophilus;* (D) polychaete worm, *Nereis.*

serve either in the uniform medium of the seawater, or in the almost equally uniform medium of a burrow in the soft mud under the sea.

Arthropods differ from polychaete worms, from which they were perhaps descended, in being adapted to walking on the substratum. This entails not only specialization of muscles but reinforcement of the body wall with chitin, which at the same time must be jointed in order to bend. A system of levers is involved in the action of every appendage. Primitively the appendages must have been of a uniform type in all segments of the body, and this type need not have been particularly different from that of the polychaete worms, except for its jointing. In fact, it may be that Crustacea come closest to showing the original form of these appendages, for primitive Crustacea have each appendage divided into an upper and lower branch, and, in addition, emerging from the base, there is a gill on the dorsal side. This resemblance to annelid parapodia may not be just a coincidence. The earliest known Crustacea, the trilobites, had lateral appendages that were very much alike along the whole body, except for mouthparts and antennae. Modern Crustacea appear to approach this serial uniformity most closely in the adults of the Branchiopoda and in some of the Malacostraca, especially the long-bodied decapods. All of these are able to use the appendages for swimming as well as walking, although there is some specialization in Malacostraca, such that the more anterior legs are for walking and the more posterior, called swimmerets, produce a current by beating against the water. Many kinds of prawns spend their time swimming freely. Among Branchiopods, the fairy shrimps, *Eubranchippus* or *Branchinecta,* swim continuously in an upside-down position. Among Amphipods the scuds, or sand fleas, such as *Gammarus,* are very rapid and effective swimmers in water when disturbed but live mostly among vegetation or feeding in sheltered places. The ventral position of the appendages on the body implies that their original adaptation among Crustacea had reference to locomotion on a substratum, but they were not sufficiently committed to that to prevent a ready availability of the same appendages for free swimming.

All of this suggests that there are two major aspects to the evolution of locomotion in the Crustacea, one being the further specialization and modification of the legs in the adults of advanced groups, and the other the very different modifications developed in the larval stages of most marine groups. The first of these does not concern us here in regard to swimming, but the second is very important because a great part of the life of the sea consists of larval, planktonic Crustacea of several different orders. In the development of a shrimp, for example, there is a *nauplius* which has three pairs of appendages. This is a microscopic, transparent, nonsegmented larva; it swims by jerking motions of these appendages; the first two pairs are antennae. This larva transforms to a *protozoea,* which continues to swim primarily by its antennae, and this in turn to a *zoea,* in which the antennae play a smaller part but the legs also become involved. Following this is a *mysis* stage, having a form of body more like that of the adult shrimp, and the thoracic and abdominal appendages are used in swimming rather than the antennae. This larva eventually changes to the adult, which is in many cases more sedentary and may not be transparent. Quite apart from the growth of the successive stages of larvae, the addition of appendages and segments, and the changes in form necessary to attain that of the adult, it is obvious that young larvae have their own special adaptations to planktonic life, and these include a very different mode of swimming from that of the large, morphologically more primitive, adult.

Swimming insects are numerous. Some aquatic beetles simply run in the water, with no specialized swimming movements or structures, but in the Dytiscidae, diving beetles, the hind legs are laterally extended, move horizontally, and are fringed with hairs so that they act as oars. These insects are large enough so that against the viscosity of the water they maintain some momentum and are therefore streamlined in shape, tapering at both ends. They are rapid and powerful swimmers, often darting up to the surface, turning to trap a bubble of air under the hind end of the wing covers, then dashing quickly to the bottom again. They also hop out of the water at night and fly, coming to lights, where they frequently are collected.

The back swimmers, Notonectidae, among Hemiptera, swim upside down, using the hind legs as oars; the shape of the body is boatlike, having a keel on the dorsal (actually the lower) side.

Some vertebrates also swim by rowing motions of paddle-like limbs. In sea turtles the front legs are much longer than the hind, flattened, and used in action comparable to the breast stroke in human swimming. Young hatchling loggerhead turtles, in their first frantic dash for the water, make energetic flailing motions that are more restricted in extent as soon as the animal enters the water but provide the propelling force in swimming. Some fishes, such as the chimaeras and certain reef-dwelling teleosts, have greatly extended pectoral fins that similarly act as the main propellers, the tail being reduced or used as a rudder.

"Flying" Swimmers (Fig. 20). It makes little difference whether we describe a penguin swimming under water as a flyer or a rower. Its locomotion (and, independently, that of another family of sea birds, the auks) is not very different from the flight of birds, bats, and pterosaurs in the air. The essential difference is, of course, in the relatively great density of the medium. This means that less surface area is required to accomplish a given amount of thrust and lift than would be necessary for the same body weight out of the water. By the same token, less vertical and horizontal wing action is needed under water, for a given amount of

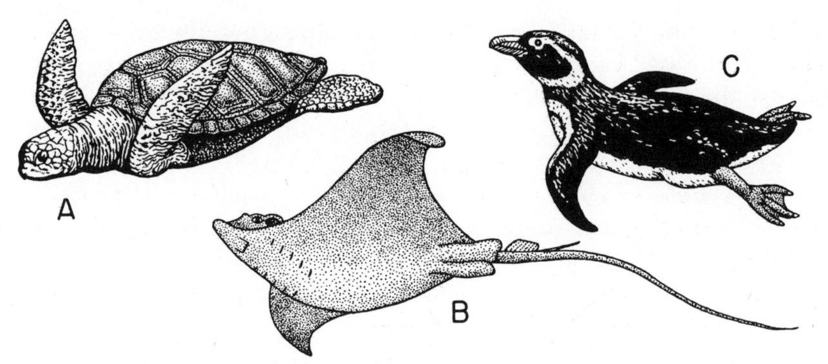

Figure 20. Animals that swim by "flying" through the water: (A) green turtle; (B) eagle ray; (C) Magellanic penguin.

effort produces a greater effect. Penguins are, therefore, very fast and agile swimmers, using their wings in this action, and are able to catch fish of considerable size by chasing them. An Adelie penguin, one of the smaller species, easily attains enough speed to shoot itself out of water to a height of 4 or 5 meters, landing on the edge of a shelf of ice.

From a quite different source, the mantas and eagle rays (related to sharks) evolved a mechanism for "flying" through the water by winglike motions of their huge pectoral fins. Most rays swim by undulations of the margins of the pectoral fins, but in the big rays the fin forms a tapering triangle and the fish flaps along lazily in the water, like a bird in slow motion.

Undulators (Fig. 21). A different kind of swimming is that in which the body undulates in waves that progress toward the posterior end. It is evidently most effective in animals of medium to large size; among the few microscopic animals that use this action are the trypanosomes, whose undulating membrane, along one edge, takes part in the motion of the flagellum that supports it. A few worms and insect larvae have a rather ineffective sort of undulation, but perhaps the best example among invertebrates is to be seen in leeches. These are ribbon-shaped, flattened below and above, and undulate in an up-and-down motion, the waves being small anteriorly but enlarging as they approach the hind end; the result is a surprisingly rapid and direct locomotion.

Most familiar of the animals that swim by wave action are the fishes and the tailed, aquatic Amphibia. Ordinarily the head is blunter, broader, and heavier than the tail end, and the waves do not begin at the anterior tip but behind the head or even farther back. In many cases the head is massive enough to act as an inertial center with respect to the contraction of muscles and the right-left movements of the body. The effective work of the body and tail against the water is increased by the expanded surfaces of median fins above and below the body, especially by a caudal fin. The waves travel back along the body and tail in such a way as to thrust against the water with the posterior aspect of each curve, causing the body to follow an undulating path through the water, especially in the more eel-like fishes.

Evolution of eel-like swimmers has taken place many times independently in fishes and Amphibia. The body lengthens, median fins extend and fuse, the tail is reduced, and the head may also be narrowed. All the known types of eel-like fishes and amphibians represent specializations from ordinary spindle-shaped or salamander-like forms.

For fast and agile swimming the opposite sort of adaptation of the normal fish body takes place, that is, concentration of movement in the tail peduncle and caudal fin. This is accomplished by bringing together posteriorly the connective tissue bands and tendons from segmental muscles along the side of the body, so that the work of swimming is done by contraction of masses of muscle and transmitted by nearly horizontal tendons to the tail. Among these fish the mackerels, tunas, and their relatives achieve the greatest speed with minimum effort; this requires a refined streamlining of the entire surface of the body, including upper and lower jaws, eyes, nostrils, gill covers, and even the dorsal and ventral contour lines of the body, and usually a deeply forked, narrow, and stiff caudal fin.

The mammalian order Cetacea, whales and porpoises, has been highly specialized for aquatic life ever since the middle Eocene at least, and therefore is so far modified in structure that a determination of its precise ancestry is very difficult. This is generally thought, however, to be among

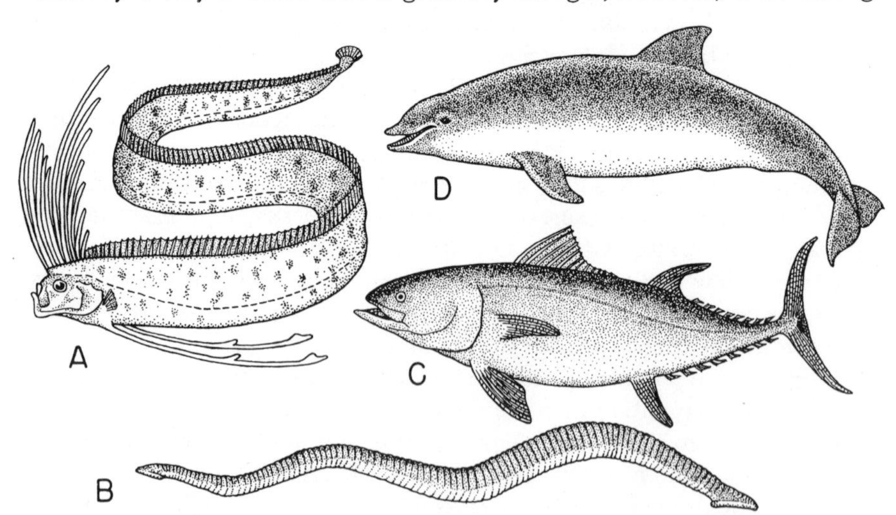

Figure 21. Animals that swim by undulating: (A) oarfish, *Regalecus glesne;* (B) leech (vertical undulation); (C) tuna; (D) porpoise, *Tursiops.*

the primitive creodonts, perhaps the Mesonychidae. One may imagine that the original semiaquatic members of the group that gave rise to Cetacea could plunge from a bank into the water, arching the body and tail, then straightening and raising the broad tail so that the animal immediately surfaced, or perhaps went through a series of arching dives. In living Cetacea there is an up-and-down undulation of both body and tail in swimming, rather than lateral. Naturally this would not be effective without developing a broad, flat lateral flange on either side of the tail, extending into powerful horizontal flukes. As is shown most beautifully in the moving picture "The Silent World" by Jacques Cousteau, whales and porpoises swing the tail gracefully up and down, along with the whole posterior part of the body. The animal may weigh from a few hundred pounds to 150 tons, and when it is under way relatively little power is needed to keep it in motion because of its momentum, but tremendous energy is required to start or to change course.

REFERENCES

Bates, H. W. 1862. Contributions to an insect fauna of the Amazon Valley. Lepidoptera: Heliconidae. *Trans. Linn. Soc., Zool. 23:* 495.

Brower, J. V. Z. 1958a. Experimental studies of mimicry in some North American butterflies. Part I. The Monarch, *Danaus plexippus,* and Viceroy, *Limenitis archippus archippus. Evolution 12:* 32–47.

———. 1958b. *Ibid.,* Part II. *Battus philenor* and *Papilio troilus, P. polyxenes* and *P. glaucus. Evolution 12:* 123–136.

———. 1958c. *Ibid.,* Part III. *Danaus gilippus berenice* and *Limenitis archippus floridensis. Evolution 12:* 273–285.

———, and L. P. Brower. 1965. Experimental studies of mimicry. 8. Further investigations of honeybees (*Apis mellifera*) and their dronefly mimics (*Eristalis* spp.). *Am. Naturalist 99:* 173–188.

Comstock, J. H. 1918. *The Wings of Insects.* Comstock. Ithaca, N.Y.

Cott, H. B. 1940. *Adaptive Coloration in Animals.* Methuen, London; Dover, New York.

Edmunds, G. F., and J. R. Traver. 1954. The flight mechanics and evolution of the wings of Ephemeroptera, with notes on the archetype insect wing. *J. Wash. Acad. Sci. 44*(12): 390–400.

Ford, E. B. 1964. *Ecological Genetics.* Wiley, New York. (Chaps. 12 and 13 on mimicry, Chap. 14 on industrial melanism.)

Müller, F. 1879. Ituna and Thyridia: a remarkable case of mimicry in butterflies. *Proc. Entomol. Soc. London 1879:* 20–29.

8 · NATURAL SELECTION AND ADAPTATION

On reading *The Origin of Species* today, a biologist may be struck by the diffidence, almost hesitation, with which Darwin put forward his arguments for the reality of natural selection, and by the seemingly superficial kinds of observations, numerous as they are, on which the idea rested. This is partially because we are accustomed to thinking in these terms and his readers were not, but even more it is due to the lack of understanding, in his day, of the nature of variation or its causes, to a considerable misunderstanding of the processes of heredity, and the absence of a body of systematic knowledge of ecology, populations, paleontology, and other fields in which Darwin himself was an early pioneer. It is an oversimplification to suggest that to Darwin natural selection meant merely "survival of the fittest" and that it means today "differential reproduction," for only a small part of the idea is conveyed in either phrase.

An example of the action of natural selection in bringing about a rapid evolutionary change in the characters of certain moths was given in Chapter 7, showing that the agency is predation by insect-eating birds, that the variation occurring in the moth population is a simple mutation in the gene for color, and that the adaptive advantage of the mutation is that its possessors survive more readily in the changed (dark) environment where they spend the day than do other individuals of the same species. Other examples were given of controlled experiments showing the reality of selective predation in the case of mimicry, indicating that predators are deceived by it in the same way that we are deceived.

Predation is only one of innumerable agents of selection. Almost any factor in the lives of organisms may serve, as long as it produces a difference between some individuals (with certain genotypes) and others (with other genotypes) in the amount of their contribution to the next generation. This means that selection results in some changes in the population during successive generations. In sexually reproducing organisms no genotype of an individual is transmitted as a whole to the offspring (this does occur in parthenogenesis and other forms of asexual reproduction), but in the production of gametes there is segregation (preceded generally by crossing over, more or less at random) and a reassortment or recombination when the egg is fertilized. This in itself is a major source of variability, and the variants are often selectable. Mutations are rare, but in the long run they provide a broadening of variability beyond the limits of reassortment.

The temptation to regard organisms as merely the genotype's way of making other genotypes is difficult for some biologists to resist. We read, for example, that natural selection is no longer considered to relate to survival but only to differences in reproductive rate. It is worth remembering that the existence (survival) of an individual from the moment of fertilization to the moment when viable offspring have been produced is essential to any contribution of genes by that individual to coming generations. Most of the events that influence its reproductive rate occur during that time. The fact that in many species the vast majority of individuals die before reproduction for reasons that have nothing to do with genetic differences is irrelevant here, for it is among the small surviving minority that the significant differential reproduction takes place, selectively, even if these number no more than 1 or 2 percent of the young of that generation. There is, moreover, a certain amount of mortality before reproduction that is due to circumstances bearing on genetic differences. It is seldom possible to determine the precise value to place on this under natural conditions, although the general effects may be obvious, as in the work of Kettlewell already presented. On the other hand, artificial selection, or a simulated "natural selection" under arbitrarily controlled laboratory conditions, is quite another question. The mathematical studies of R. A. Fisher, S. Wright and J. B. S. Haldane demonstrated during the 1930s that natural selection is effective in populations even when the difference in adaptive (reproductive) value of the characters concerned is 1 percent or less. That is, suppose that a particular allele confers a probability that 100 eggs will hatch, compared to every 99 for its alternative, the former, in time, spreads through the population. Actually many genetic alternatives are now thought to differ by much larger adaptive value than this, and their spread is correspondingly more rapid.

This chapter will be given to discussion of several phenomena or ideas relating to adaptation and selection which are currently being studied and on which there are differences of opinion, especially as to what the evidence actually means.

GENETIC ASSIMILATION, OR THE "BALDWIN EFFECT"

Structural or functional adaptations may develop in an individual by the direct response of tissues to environmental stimuli or to physiological stimuli in the body. For instance, the strength, size, and facility of action of muscles increases as a result of exercise, and this may be limited to certain muscles that are used while others are not. One kidney, if the other has been removed, increases in size and in its capacity to handle the complex labor of excretion. The soles of the feet become thick and tough as an effect of walking. But these adaptations are "exogenous" or physiological, and nothing is inherited except the ability to make such responses.

Then how does it happen that an unborn fetus has soles that have already begun to thicken? How do the pectoral muscles and associated skeletal parts of a mole become heavy and powerful, or start to do so, before being used? How can an ostrich, on hatching from the egg, show callosities on its belly and breast such as it would receive from resting on them later? At first glance this looks like evidence for the inheritance of acquired characters, for there is no doubt that these examples are genetically determined. Because of their resemblance to the nonhereditary traits mentioned first, they are called "pseudoexogenous."

It was suggested by Mark Baldwin in 1896, in the *American Naturalist*, that hereditary variations that happened to coincide with acquired modifications would be favored by natural selection and would be sustained until they became fully established in a species. C. Lloyd Morgan (1900, Animal Behaviour) accepted this explanation, adding: "Survival would in the long run be better secured, we may suppose, where the two methods of adjustment were coincident and not conflicting." Morgan applied the idea especially to the evolution of animal behavior in an attempt to explain the origin of instincts by fixation of acquired habits in the genetic constitution of a species. "There would be a distinct advantage in the struggle for existence when inherited tendencies of independent origin coincided in direction with acquired modifications of behaviour."

The "Baldwin effect" thus implies that nonhereditary adaptations, including those of behavior, provide a means of sustaining a species during such time as may be required for mutations to appear and to spread through the population, causing the same adaptation to have a genetic basis. Presumably at this point direct environmental control might be relinquished, or habit might become instinct. The question now seems to be whether it is necessary to postulate two different causes for a particular adaptation, one gradually substituting for the other, and whether, if this is so, selection of mutant genes leading to genetic reinforcement of a character already produced in another way is probable, or even possible.

One has to imagine, first, gene mutations that just happen to cause an animal to have the same pattern of behavior, instinctively, that its species already achieved by learning, or calluses in precisely the same area of skin where they already appeared as a result of walking. One must then imagine selection favoring those individuals that are so equipped, even though they and all others normally develop the same behavior or structure by (1) learning, or (2) rubbing, as the case may be. This seems highly improbable.

Waddington (1957, p. 141) discussed the problem in relation to natural selection, proposing ways in which exogenous adaptive characters might be "assimilated" into the genetic constitution of a species. Granting that a species inherits the ability to respond to an environmental influence by producing an adaptive character, he proposes that selection will "mould the epigenetic landscape" (that is, the processes of development) into a new form, which increases the organism's ability to reach a favorable result, perhaps by means of genes that switch development into an easier pathway. He cites several experiments to indicate how this takes place. One is quoted here:

> These experiments were made with *Drosophila*. Very strong environmental stimuli were used, which pushed development over well-marked thresholds into quite definitely abnormal channels. In a first series of experiments, pupae aged about 21 to 23 hours were subjected to a temperature of 40°C. for four hours. In the foundation stock, a number of aberrations in the wings were produced. One of these, a breaking or even complete absence of the posterior crossvein, was selected for study. Selection was applied for (and also against) the capacity to react to the environment in this manner, the "upward" selected stock being carried on by breeding in every generation from flies in which the crossvein was broken, while the "downward" selected were bred by taking in every generation those which failed to respond. It was immediately apparent that, as might be expected, the capacity to respond was under genetic control and became strengthened (or weakened, as the case might be) as the experiment proceeded. The important point then emerged that genetic assimilation began to occur. After about 14 generations, flies of the upward selected stock were found to produce a small number of offspring which developed broken crossveins even when they were not given the temperature treatment. In order to speed up the further process of assimilation, these flies were bred from and selected in normal temperature, when stocks were rapidly produced which had a broken or absent crossvein in a high percentage of individuals.

In these experiments there was no ready-made adaptive character, but an abnormality that appeared as the occasional effect of an unusual environmental stimulus, heat. But from the beginning the capacity to make this response was under genetic control; that is, the genotype permitted it. Selection during 14 generations of flies increased the frequency of the response and also extended the range of the temperature stimulus to which response was possible until it finally included "normal" temperature. The range of sensitivity of the pupae had changed. After this was done, selection for the character "broken or absent crossvein" could, of course, be carried on without using the heat stimulus. One might just as well say that normal temperature was itself the stimulus. But there is no evidence that any mutations occurred that just happened to produce exactly the same effect, broken or absent crossvein, that was at first produced by heat. The "genetic control" of this character does not imply that it was a result of a mutation in the first place.To say that the character has been assimilated into the genetic composition of the species is to suggest far more than the evidence permits. Actually selection seems to have modified the responsiveness of the organism to one factor, temperature, at a stage in the development of the phenotype; this is, in effect, a change in the expression of the gene or genes that permitted the abnormality to occur under higher temperature in the beginning of these experiments. It is perhaps comparable to changes in the expression of the gene for "eyeless" because of selection, as shown long ago by Morgan (1929).

The same idea may fit other instances of the "Baldwin effect" without assuming new mutations to fix, genetically, a character that was produced "by the environment" initially. A feature that is in some measure adaptive, but that does not appear unless the organism interacts with the environment in certain ways, is not on that account lacking in a genetic basis. We can say that a genotype exists which, given a certain environmental stimulus, is capable of producing the adaptive phenotype, for example, calluses. If such a genotype were not, at first, common to the whole population, or only occasionally permitted this particular response, then selection obviously could (1) increase its frequency in the population and (2) alter its expression by favoring a greater sensitivity or responsiveness to a wider range of stimuli. Environmental stimuli are rarely discrete units or unvarying quantities, and neither are the phenotypic expressions of genes.

GROUP SELECTION AND INDIVIDUAL SELECTION

Although populations, not individuals, evolve, the only method by which this is accomplished, as far as we have seen to this point, is the differential survival and reproduction of individuals, resulting in the slow shifting of gene frequencies in a population. But is it not possible that some populations in a species are better adapted and more likely to persist than others? May not some characters be significant to a population as a whole rather than to individuals, and therefore be explained better by differential survival of groups? If so, they should be either neutral or deleterious to individuals and thus inexplicable by individual selection but of obvious survival value to a group or population. They must, of course, be hereditary and transmissible in sexual reproduction.

At first thought, a family group among birds or mammals qualifies as one in which some characteristics have great value to the group but none to the individuals that possess them. The complex nesting behavior of most birds, their expenditure of time and energy in labor that has value only to the eggs and dependent young, their exposure to predators, often in a way that looks like self-sacrifice, their cooperation with each other, and so on—all this would seem to have evolved as a burden to be borne by reproducing individuals for the benefit of a group. But in terms of natural selection, it must be remembered that the young are carrying the genetic contribution of the parents and that the business of reproduction is not finished when the eggs are laid; it is not finished until the nestlings have become independent. Thus the production and survival of this particular contribution to the gene pool is still based on characters evolved to serve individuals, to make possible their reproduction. It is not necessary to call on differential survival of groups to account for any of the characters concerned in this, even though it is true that there is differential mortality among such family groups. As far as this may be genetically based, it affects the reproductive rate of the parent individuals.

Letting the group be a deme (or local population) or a herd or school in which the offspring of certain parents are not the important element, we meet difficulties of another kind. How is the relative success of different populations to be decided, other than by the extinction of some and persistence of others? How is it determined that a genetic characteristic is responsible for survival versus extinction and that it consequently changes in the course of time?

Many biologists have tried to establish criteria of population success, variously favoring total number of individuals, rate of change in number, total mass, length of survival of the population, ecological versatility, stability of numbers, and so on. Williams (1966), discussing this at length, accepts the last of these. But there are many kinds of animals and plants in which a rhythmic fluctuation in numbers is inherent in their ecology, as, for instance, the numerical relationship between insect parasites and their hosts, yet could scarcely be taken to show differential success in an evolutionary sense. If we are discussing the possible selective value of differences between populations in the same or closely related species, such as would produce evolutionary change, it seems to the writer that no single criterion among those listed, or any other he can think of, could be applied very widely. Extinction settles the matter for many populations, but extinction does not, even indirectly, cause evolution.

Another point of difficulty is that most observable natural populations do not come and go, appear and disappear, in a year or two, but may persist for centuries or millenia without conspicuous change, without splitting, fusing with others, or facing radically different circumstances. But new generations arrive annually or more often in most species, and the individual variability that comes up for selection is acted on at a rate enormously more rapid than is conceivable for populations. If we then attempt to think of selection in terms of small parts of populations in order to make the question more manageable, we come back again to individual selection.

There is, of course, ecological replacement. If three groups out of four that were present at a certain time in geologic history become extinct, then the fourth is the survivor, and it may strongly influence the composition of the biota. Ecological dominance by certain groups has sometimes been overthrown by other groups. For instance, the big stereopondyl amphibians of the Triassic were apparently crowded out, perhaps exterminated, by the phytosaurs, and these in turn were replaced rather promptly in the same ecological situation by crocodiles. But this is not an example of group selection. These animals were not related to each other; no genetic change was accomplished. The characteristics that caused them to occupy similar ecological niches in succession were not due to a sequence of adaptations in a single phylogenetic line. They were faunal substitutions based on somewhat parallel, independent adaptations, and these can be explained by ordinary individual variation and selection.

An example of supposed group selection was mentioned by Lewontin (1966) in his review of the book by Williams (1966). It concerns "altruistic" behavior for the benefit of a group such as a flock or a herd, not necessarily immediate relatives, that is detrimental to the individual doing it.

> For example, if a bird gives a warning cry or makes a conspicuous display of its plumage at the approach of an enemy, all the individuals in the population will be benefited by the warning, but there is no special differential benefit in fitness to the altruist who has given the warning. On the contrary, the very act of giving a warning cry or making a display will call the bird to the attention of the enemy and in this way may prejudice the survival of the foolish Samaritan. Since any genes causing such behavior will not be increased in the population by differential fitness and, indeed, may be decreased, we cannot explain the evolution of such characters by Darwinian selection.

If one answers this by saying that the action contributes to the survival of a genotype that is probably common to the members of the group and is transmitted by them even if the "altruist" did not survive, then I think he has overlooked an important fact. He assumed that the warning cry or display puts the individual giving it in danger by calling him to the attention of the enemy. So it would, if the enemy carried a gun, but otherwise any bird that has noticed an enemy and is sufficiently alert to give a warning has already put itself out of reach of a natural predator and is safer than other members of the flock, which had not been aware of the enemy's approach. It can well afford to call attention to itself and does not thereby prejudice its chances of contributing to the next generation. If a predator comes close enough to be inescapable before it is noticed by any in the flock, the victim usually has no chance to give a warning.

The ability of other members of the prey species to react immediately to the cry or display, or to sight of a predator, can readily be explained by individual selection. To speak of a "warning" as an example of altruism for the benefit of the group is to account for it in human terms; we do not know that the motives of birds are the same as ours would be, and in any case the motive is irrelevant. Explanation by means of group selection is unnecessary, for the "altruist" as well as the group is benefited by exposure of the enemy to the attention of all. I agree with Williams that other cases of apparent cooperation among animals do not, in general, require group selection for their development.

One additional question should be mentioned as having given some difficulty for the theory of individual selection, although it was considered and solved by Darwin. Complex societies have evolved in insects, once in the ancestry of termites (Isoptera), several times in Hymenoptera, and to a small extent in certain other groups. Among termites the colony contains sterile, wingless workers of both sexes, with one or more specialized castes (as soldiers, nasutes, and so on), and a small number of reproductives. All are diploid in chromosome number. Commonly one male and female, after a mating flight, establish the colony, and thereafter the queen, wingless and much enlarged, remains in a chamber attended by a male and by workers, producing great numbers of eggs during a life of several years. In Hymenoptera the workers may or may not show a caste system, and are sterile females only. Males are haploid. There may be one or several queens, depending on the group, and they are not attended by males after the mating flight, but sometimes the female mates with more than one male during the flight. The question, under these varied conditions, is whether group selection must be invoked to account for the complicated adaptations of these societies, especially in view of the fact that reproduction is often limited to less than 1 percent of the members of a colony.

Naturally it is not difficult to find examples of beehives, wasp nests, or ant colonies that persist when others fail, grow larger while others dwindle, or outlast their neighbors. But exactly what is being selected in this case? Not a population of many thousand individuals whose genotypes collectively will be transmitted, but one reproductive female and her sterile offspring, plus potentially a few reproductives that can start other colonies. The effective breeding population in one colony is the same, usually, as in a pair of nesting birds, a male and a female. The multitude of nonreproductive workers contribute nothing except to maintain a breeding facility for that pair; the colony is, in a sense, a "superorganism" built by and around the queen. The coming generation, as far as evolution is concerned, is the reproductive minority of this queen's offspring. Whether they appear at all, and how successful they may be in founding new colonies, are the criteria of selection, which thus has an individual genotypic basis.

EVOLUTIONARY PLASTICITY AS AN ADAPTATION

The principal source of genetic variability, as already noted, is the random assortment and recombination of genes in sexual reproduction. As there may be thousands of gene loci involved in this shuffling, the number of different combinations that are possible is astronomical, particularly because of the interaction and interdependence of genes in producing their effects. Inasmuch as sexuality makes this enormous plasticity available as the material for natural selection, it is not surprising that some authors (for example, Huxley, 1958; Darlington, 1958) suggest that sex was an adaptation to provide variability in the first place. As the mutation of genes further extends the range of variability in the course of time, it has also been proposed that the occurrence and rate of mutation is somehow an adaptation brought about by selection to promote plasticity, in order that the chance of adaptation to future circumstances shall be improved. Finally, natural selection itself has been conceived by some as an adaptation to promote evolution! We may consider these ideas briefly, but for a more extensive comment see Williams (1966).

Natural selection is not a regulatory mechanism imposed upon populations from the outside to ensure that they shall evolve. It is simply the history of individuals and populations in terms of survival and reproduction; it is what they are doing and have done. Change is not inherent in this history. The elimination of maladaptive variations as they come along is one form of selection, the result of which may be to maintain the characters of a population in status quo. Indeed, under most conditions the greatest contribution to coming generations is made by individuals with characters and genotypes at or near the mean of variation in their population. If circumstances change, then adaptive values also, in part, change, and selection then can bring the genotypes and phenotypes to another status. But there is no mystical power of selection to lead to higher things, to promote progress, to manage, beneficently, the advance of life.

The idea that the rate of mutation can be adjusted by selection to assure optimum adaptability against future conditions, not yet faced, would seem to require a wholly unscientific mysticism. Selection affects what exists. For events of future years or future centuries to be anticipated in the differential reproduction of current organisms is unthinkable. The same criticism applies to the notion that sex was an adaptation to promote variability, supposedly for the sake of the evolution of generations yet unborn. Who or what foresees that there might be an advantage in variability and arranges the appropriate selection to bring it about? The conjugation of gametes could cause diversification of genotypes in the ensuing generation, as compared with the constancy of those produced by nonsexual spores; under especially stringent circumstances favorable phenotypes differing from those of the parents might sometimes occur and maintain themselves, but it does not seem reasonable to admit any teleology into modern biological thinking. The writer can only conclude that there is no selection for the sake of evolutionary plasticity as such, and that the occurrence and frequency of mutations cannot possibly be, in themselves, adaptive.

PREADAPTATION

When a new taxonomic group of distinctive character arises, it commonly does so by a major shift in adaptation, entering a different ecological setting or a new "adaptive zone." If we happen to know the direct ancestors of the new taxon we can then recognize which adaptive characters of the ancestral stock were inherited by, and indeed made possible, the new.

For instance, certain sprightly bipedal reptiles of the order Thecodontia were undoubtedly the ancestors of birds. A few of them were almost certainly arboreal, and the combination of bipedal habit, light build, and agility, as inferred from the skeletons, were passed on with virtually no change to the Jurassic bird *Archaeopteryx*. In the latter, if it were not for the feathers, we should have no reason to recognize a new class of vertebrates or even a new order of reptiles. In retrospect we can say that some thecodonts were "preadapted" in certain ways for the life of birds. But it is all too easy to take another step in thinking and imagine that thecodont reptiles existed during the Triassic in actual anticipation of birds, as if the term "preadaptation" meant that the birds were foreseen as the purpose of being thecodonts. This is teleology and has no justification whatever. If we allow "preadaptation" to have such a meaning, then the word is worse than useless. Adaptations undoubtedly exist in many if not all groups of organisms that could in some way make possible an additional step into a new adaptive zone, but ordinarily the occasion for making such an advance does not arise. The "pre"-adaptation is known as such only when seen in retrospect.

As another example, the Devonian freshwater lobefinned fishes were fishes purely and simply. There could be no reason for an ichthyologist, if he were carried back to a Devonian swamp, to suspect that in them lay a hidden purpose of producing Amphibia and ultimately the reptiles, birds, and mammals. Their adaptive fish characters included a pair of lungs, peculiar paddle-like pectoral and pelvic fins that could with little modification give rise to legs and feet, a pair of spiracles that, becoming closed externally, made the middle-ear cavities of tetrapods, and so on. To the extent that we are able to point out these homologies with something characteristic of tetrapods, it is possible to say that the lobefins were preadapted as amphibians. But the term seems entirely unnecessary. It does not represent a real phenomenon, and it carries a false implication that we must be at pains to disavow, for ancestors have their own justification for living quite apart from any possible descendants.

REFERENCES

Darlington, C. D. 1958. *The Evolution of Genetic Systems.* Basic Books, New York.

Fisher, R. A. 1958. *The Genetical Theory of Natural Selection,* 2nd ed. Dover, New York.

Haldane, J. B. S. 1932. The time of action of genes and its bearing on some evolutionary problems. *Am. Naturalist. 66*: 5.

Huxley, J. S. 1958. Cultural process and evolution, pp. 437–454. In A. Roe and G. G. Simpson (eds.), *Evolution and Behavior.* Yale Univ. Press, New Haven.

Lewontin, R. C. 1966. Adaptation and natural selection. *Science 251*: 338–339.

Morgan, T. H. 1929. The variability of eyeless. *Publ. Carnegie Inst. 399*: 139–168.

Waddington, C. H. 1957. *The Strategy of the Genes.* G. Allen, London.

Williams, G. C. 1966. *Adaptation and Natural Selection, a Critique of Some Current Evolutionary Thought.* Princeton Univ. Press, Princeton, N. J.

9 · DEVELOPMENT AND EVOLUTION

Preevolutionary classifications of animals and plants resulted in a linear scheme, in which man was at the pinnacle and other organisms arranged in descending order of complexity below him. This, of course, was difficult to achieve, except in a rough way and by the exercise of considerable imagination, because distantly related groups (as we see them now) may not fit logically into a single sequence. After the theory of evolution became widely accepted through the work of Darwin, this "scale of beings," with its implication of upward progress, seemed ready-made to fit it, provided only that the scale were pictured as a branching tree rather than a single line.

Since the invention of the microscope another scale has appeared in the thinking of biologists: that of individual development (ontogeny) from egg to adult. Much thought has been given to relationships, if any, between these two scales and the possibility that one might be explained by the other. Von Baer (1828) concluded that:

1. General characters of animals appear earlier in development than special characters.
2. Special characters develop from the general.
3. Different kinds of animals are more alike in early, but less alike in late, stages of development.
4. Young stages of an animal are not like adults of other animals lower in the scale but like young of those animals.

Von Baer's "laws" are not natural laws in the sense of cause-and-effect relationships, nor are they meant to explain, but they state the commonly observed facts. Today they appear an oversimplification, with many exceptions, but they are useful to any student of biology as a guide in his thinking. It was apparent to Darwin and to students of evolution after him that the reason for the resemblance of early stages of animals in which adults differed was their descent from a common ancestry, although this need not imply that young were comparable to adult ancestors.

Haeckel (1866, 1874) drew exactly that implication, however, holding that development is a condensed version of the evolutionary succession of adult ancestors; that is, the scale of ontogeny simply recapitulates, briefly, the scale of phylogeny. This suggested that somehow, through the mechanism of heredity, phylogeny must be the cause of ontogeny. Haeckel found examples of recapitulation among the lower invertebrates particularly. He thought that single-celled Protozoa gave rise to simple multicellular animals by remaining attached after fission and forming a layer of cells around a cavity, as in the planula larva of a jellyfish. Adult coelenterates of simple form, having an inner layer of cells that invaginated from the outside and consequently a mouth, were thought to have preceded the three-layered animals, illustrated first by flatworms.

Haeckel compared these supposed stages of evolution with the early stages of embryology of most higher animals, in which a cell (the egg) becomes a blastula (cells enclosing a cavity), then by invagination a gastrula of two layers, ectoderm and endoderm, after which the third layer, mesoderm, develops between. (For an extensive discussion of the origin of Metazoa, and current differences of opinion, see Dougherty et al., 1963.)

Another example of supposed recapitulation is the parallelism between a series of developmental stages of some Crustacea (for example, a prawn, *Penaeus*) and a series of adults of certain selected crustacean orders. The eggs of this marine prawn hatch into minute transparent nauplius larvae, unsegmented but with three pairs of appendages (the first and second pairs of antennae and the mandibles). This swimming larva molts, becoming a metanauplius, then in succession a protozoea, zoea, mysis, and finally an adult, in each stage adding more segments and more pairs of appendages. Each of the larval stages resembles rather closely the adult of one or more other crustacean orders (prawns are Decapoda). The nauplius has often been compared to a copepod, and as the number of segments and appendages increases, the later larvae resemble various other orders; the "mysis larva" is like the genus *Mysis* in the order Mysidacea, and so on. This has suggested that the sequence of *developmental* stages of prawns, shrimps, and their relatives was a repetition of *evolutionary* stages in the class Crustacea, on the assumption that evolution had proceeded by adding the later developmental stages, one at a time, to those that were already present in the ancestors. If true, this meant that the most advanced or specialized Crustacea were those with the largest number of larval stages, and, incidentally, of segments and appendages, whereas the primitive crustacean ancestor was probably a minute, nonsegmented creature with no more than two or three pairs of appendages. (See, for example, MacBride, 1914.)

But if the ancestral crustaceans were like the nauplius larva, it becomes nearly impossible to imagine any relationship of Crustacea with any other animals, and it would be a strange coincidence indeed that the most "advanced" orders were those in which we do see most clearly such relationships in segmentation, appendages, internal anatomy, and so forth. In fact, it is only by turning the "recapitulation" sequence upside down that we come to any reasonable understanding of evolution in this order. Beginning with animals that have many segments, many pairs of appendages (more or less alike in their function), and a sequence of several larval stages, we can arrive at the minute, abbreviated, and very specialized types by removing, one at a time, the latest stages in the life history, each time permitting reproductive maturity earlier, until what were originally early larvae become adults themselves and reproduce with great rapidity in the sea or even in temporary bodies of standing water. This is evolution promoted by *neoteny*, by simplification from a more complex ancestor, but it has nothing to do with recapitulation.

The work of numerous embryologists and paleontologists during the last hundred years demonstrates that the theory of recapitulation seldom fits the known facts of embryology and evolution. This does not mean that ontogeny has no relationship to phylogeny. There are many and diverse relationships that cannot be expressed by simple rules.

Some general features of development should be noted here. Processes of organ formation and the building of an embryo, larva, and adult are carried on by two sets of factors: those that are environmental (water, food, temperature, and the physical and chemical nature of the medium) and others that are internal (the inherited genetic endowment and actions directly initiated by it). The continuous mutual dependence of heredity and environment during development can easily be seen by the effects that result from varying either one experimentally. The characters of organisms develop in part because of the materials supplied and in part because of the genetically determined ability of the organism at particular stages to respond to these materials and to be influenced by conditions. Genes cannot alone produce parts or characteristics of organisms, but they influence the handling of materials by which those are made. Evolution, then, consists of changes in the inherited capacity of organisms to respond to conditions and to use materials in any stage of ontogeny.

Another general concept related to this is that different ontogenetic stages in one species are adapted, by selection acting upon genetic variations, to the circumstances of their lives, whether these are the same as or unlike earlier or later stages. Thus larvae differ from adults usually because they are living in very different conditions and are adapted to these. Sometimes there are several successive and dissimilar larval stages, as already seen in Crustacea and illustrated here in one species of fish, the angler, *Lophius* (Fig. 22). Its youngest larva is much like that of most marine fishes: small, slender, and transparent. Later larvae develop pigment, live among drifting seaweed, and undergo radical changes of color and shape, being concealed from predators by resemblance to their greenish-brown background. The adult, finally, changes again for a new situation resting motionless on the bottom and waiting for its prey to come close, attracted by the moving lure on the head of the angler fish.

It has been known for many years (Haldane, 1932) that effects of genes may be seen in all stages of development—gametes, fertilized eggs, early and late embryos, larvae, postlarval young, adults, certain characters of the latter associated with the production of offspring, and even adaptive characters that appear in dead parts only (for example, the pappus of a

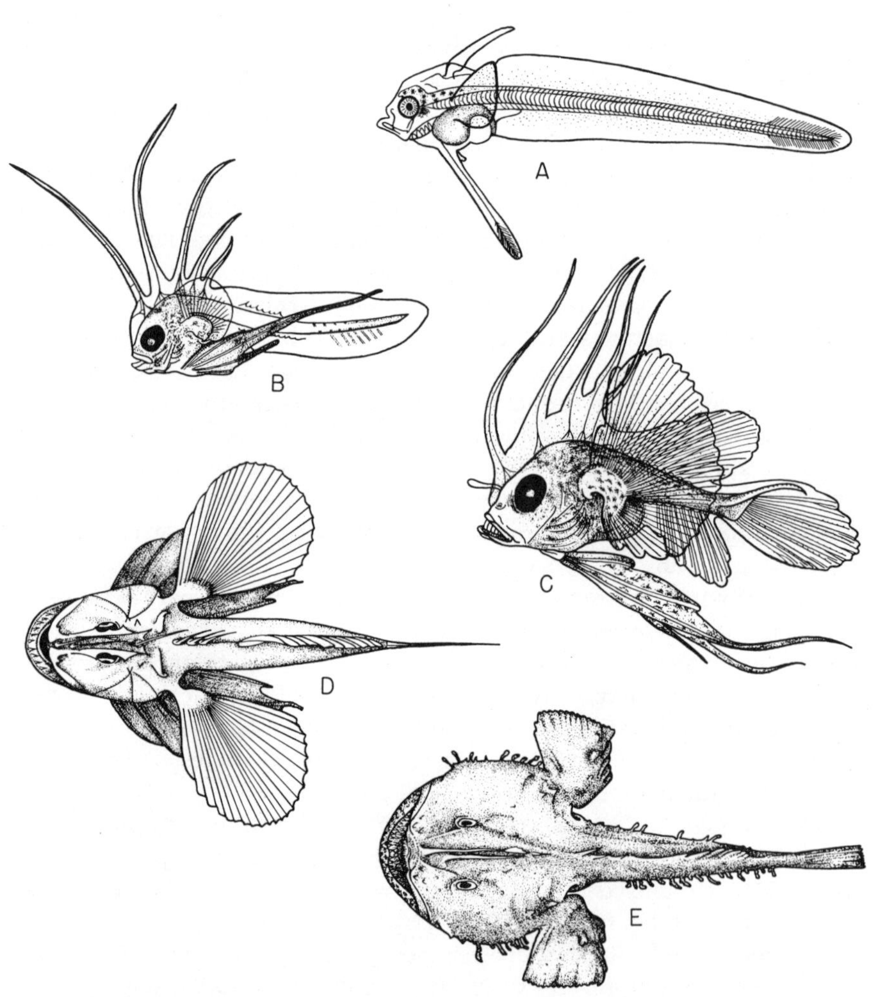

Figure 22. Transformation of the angler fish, *Lophius*, during its development: (A) 5 mm; (B) 12 mm; (C) 26 mm; (D) 52 mm; (E) adult.

dandelion, hair of mammals, and feathers of birds). This merely demonstrates that various components of the genotype can affect characters of an individual at any stage of ontogeny. Related to this is the fact that a given characteristic may appear earlier and more rapidly (or later and more slowly) in one individual than it does in another of the same species and that these differences often have a genetic basis. (See DeBeer, 1958, for a stimulating discussion of this subject.)

It would be convenient to have a simple method for expressing the rate of development so that differences between species or between individuals could be stated. Of course, this can easily be done for certain features, such as total size (length and weight) attained in a given time, time for reaching sexual maturity, or rate of growth of a particular part or dimension. But neither growth nor development (in the sense of differentiation of parts) is uniform for all parts and all functions. The shapes of adult organisms are attained by differential growth of parts; a twig of a tree grows far more rapidly in one direction than in any other, and so does an antenna of a cricket. Most leaves of trees grow in two directions (although more rapidly in one of these) but scarcely grow at all in the third. These differences in growth rate are also found in the internal organs of animal bodies. Specific patterns of differential growth change with time, not being maintained throughout life. For example, the tail of a frog starts its growth before hatching, increases rapidly thereafter, then slows, and finally stops. This organ atrophies at the end of the tadpole stage, but in the meantime the hind legs make rapid growth in the late tadpole and continue to grow after metamorphosis. The front legs appear a bit later, grow less rapidly, and retain a smaller size than the hind legs. The initiation of bone in the skeleton may not occur for weeks, months, or even one or two years after hatching, depending on the species of frog, but at, and following, metamorphosis the growth of bone is rapid, gradually becoming slower until the frog reaches its maximum size (that is, throughout its adult life).

It would be impossible, then, to express "rate of development" for an animal as a whole. Given similar circumstances, individuals of a species usually develop in much the same manner and accomplish essentially the same results in a given time. But differences do occur in the genetically controlled growth rate of particular parts or of the development of various nonlinear characters, such that the shape, total size, or proportions of parts of individuals vary noticeably. It is but a step to the equally obvious fact that these features may differ among races in a species and among species in a genus. As an illustration of the first, contrast pygmies of the Ituri Forest with the Watusi of the highlands of East Africa; for the second, compare species within almost any large genus of animals or plants (the giant toad, *Bufo marinus,* is several times as large as the oak toad, *B. quercicus*). These differences are genetic, not due to food or environment.

Many examples of differences in the rate of development of nondimensional characters have been studied. The difference between a fully pigmented goldfish and one that is nearly white is due to a single pair of alleles. One of these, when homozygous, causes a normal rate of pigment deposition, the other a rate so slow that very little is deposited in the life of the fish. The heterozygous genotype gives a rate of pigmentation that is intermediate between the other two, and therefore a color that is pale but not without gold pigment. Similarly, in the common marine crustacean, *Gammarus,* one genotype produces red eyes that are only slowly changed to black by the deposit of another pigment, but in the contrasting genotype the eye color quickly becomes black.

In the gypsy moth, *Lymantria,* studied by Goldschmidt (1927, 1934) the physiological and structural differences between males and females were shown to arise from the interaction of genes that initiated male-producing or female-producing processes in a seemingly competitive way. By crossing races that differed in the rates of these developmental effects in one sex or both, it was possible to bring about reversal of sex in some individuals and a large series of variants from extreme males to normal males, intersexes, normal females, and extreme females. Goldschmidt demonstrated that these effects resulted from control of the rate of development of male and female characteristics genetically.

The genetic effect upon the rate of a developmental process is commonly referred to as if it were the rate of action of the gene (that is, a fast-acting gene has one effect, a slower acting gene has another). But in complex multicellular animals, at least, there is probably in most cases an extended series of physiological processes leading to the observed results rather than direct action (whatever that might be) of genes. We know what some of the steps are in, for example, the action of hormones in vertebrates. Hormones that induce the formation of primary and secondary sexual characters are carrying out the dictation of the genes that determined sex in the fertilized egg. But as these hormones do not exist until endocrine tissues have developed which can produce them, obviously several processes must precede the formation of and response to sex hormones. The determination of adult sexual characters in gypsy moths and other insects is not carried out in the same way (as is shown by occasional gynandromorphs that are male on one side and female on the other), but is probably the result of a chain of events following the original genetic determination nevertheless.

Just as for any genetically controlled characters, there may be various immediate or direct causes of the rates of particular ontogenetic processes, even though the initiation of the processes was genetic. A case commonly cited is the red flowers of primroses grown at 20°C, which are white if grown at 30°C, even though the genotype is the same; temperature affects the rate of the process. Huxley (1923) showed that the rate of development of the thyroid gland (hence of the secretion of thyroxin) was correlated with differences in times of metamorphosis in several species of frogs, but these differences are themselves adaptive and have a genetic basis. Thus the causes of metamorphosis, and the transformations involved, are intermediary processes. Mutations in these as in other features provide the original material for modification of adaptive characters and for the action of natural selection.

Neoteny is the delay in development of some, or most, characters of an organism relative to its sexual maturity. This is seen in the life histories of certain salamanders in which metamorphosis and emergence from the water may not take place. In one widespread North American genus, *Ambystoma,* neoteny is characteristic of populations living in isolated pools in the arid southwest and in the Mexican plateau, where a persistent aquatic larva called "axolotl" becomes sexually mature and lays eggs but merely grows to a large size in the water and does not lose its gills or otherwise transform. We can make it do so, however, even at a fairly small size, by administering thyroxin, the hormone of the thyroid gland. Similar neoteny occurs in some individuals of the related big salamander of the Pacific coast redwood forests, *Dicamptodon,* and occasionally in the eastern spotted newt, *Diemyctylus.* In another genus, *Eurycea,* three species in Texas and Oklahoma are normally neotenous, but other, eastern, species are not. Two kinds of white, blind salamanders living in the total darkness of underground streams in Texas and Georgia are permanently neotenous.

In addition, certain members of four families of salamanders, including *Proteus* of Europe, *Necturus* and *Cryptobranchus* of the United States, and others, never transform and cannot be induced to do so by thyroxin. Their neotenic history is probably very ancient, and they may represent different stocks thus modified from an ancestral source in which the adult was terrestrial. As *Necturus* and the axolotl are common laboratory animals, and as the former, especially, has often been presented as a "primitive amphibian," it is important to point out the change in thinking on this point that has resulted from increased knowledge of fossil Amphibia. The more fully ossified skulls and skeletons of terrestrial adults in several groups (especially Salamandridae and Hynobiidae) are most readily comparable to those of early Amphibia of the Pennsylvanian and Permian periods. Larval (and neotenic) skeletons are not only less fully ossified but lack a number of *primitive* features that appear at transformation. Larval characteristics are simply those of immature, not primitive, animals. Therefore *Necturus* does not show the features of primitive Amphibia, but those of larvae, and its skeleton is specialized by reduction.

The reason neoteny has received particular attention in these animals is, in part, that it emphasizes the contrast between "adults" and "larvae," and in part that it increases the difficulty of deciding, in many cases, between those characters that are actually primitive and those that are merely larval. The term "neoteny" is commonly defined as the failure of a larva to metamorphose while becoming sexually mature. But this is a matter of degree. There are countless examples among animals (some of them mentioned below) of slight, moderate, or extensive retardation of the development of "adult" features, whether metamorphosis is or is not present in related kinds (see DeBeer, 1958). It seems better to broaden the scope of the definition than to restrict it to only a part of the wide range of these manifestations.

A neotenic effect sometimes results from an actual shortening of the time required to reach sexual maturity, as if this function had been accelerated without corresponding acceleration of other aspects of development. This accounts for the peculiarities of the pygmy sunfishes, *Elassoma* (Fig. 23), in which at maturity the body has not yet taken on the deep, rounded outline of related fishes, the bones of the cranium have not wholly ossified, the cheekbones and lateral-line canals of the head are incomplete, the lateral line on the body is absent, and the number of fin rays is less than in other sunfishes. In the hurry to become adult, the final (and primitive) characters of most adult sunfishes have been eliminated from the life cycle, and *Elassoma* reaches little more than an inch in length.

Similar results, without reduction of size or shortening of the time of development, also occur when the primitive rate of development of certain organs or systems is retarded, so that changes that occurred ancestrally are no longer possible, even though the organs in question are retained. The skeleton of vertebrates is in part developed first as cartilage, and usually this is replaced by bone (with the addition of dermal bone as well). But in sharks cartilage is simply retained as the skeleton of adults, and the capacity to replace it or to develop any kind of bone has been lacking since the Devonian period. This is a specialization, and the shark skeleton

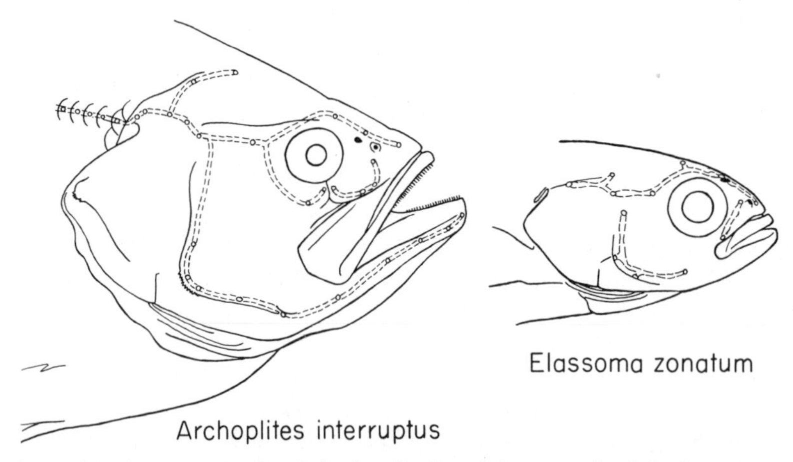

Figure 23. Sensory canals of the head of sunfishes: on the left the primitive arrangement in Sacramento perch; at right the incomplete (reduced) pattern of the pygmy sunfish. (Not to scale.)

is therefore by no means representative of the skeletons of primitive vertebrates. Another example is the persistence of down feathers throughout life in the ostrich and other ratite birds, which have probably been flightless for many millions of years. The usual flat and firm vane seen in both contour and flight feathers of flying birds *after* the nestling stage has been abandoned.

Bolk (1926) and others since have noted that many features of adult man resemble those of immature or fetal apes and even monkeys (Fig 24). The word "fetalization" does not explain how these resemblances happened, nor is it the name of a cause of them. One might not be able to argue that a particular process was responsible. Nevertheless, the scantiness of hair on the human body can properly be taken as the result of a delay in the normal process by which mammals produce hair. The light color of the skin of certain human beings is due to failure (or reduction) in the usual process of developing pigment. The position of the big toe parallel to the others is due to retaining its early embryonic position in-

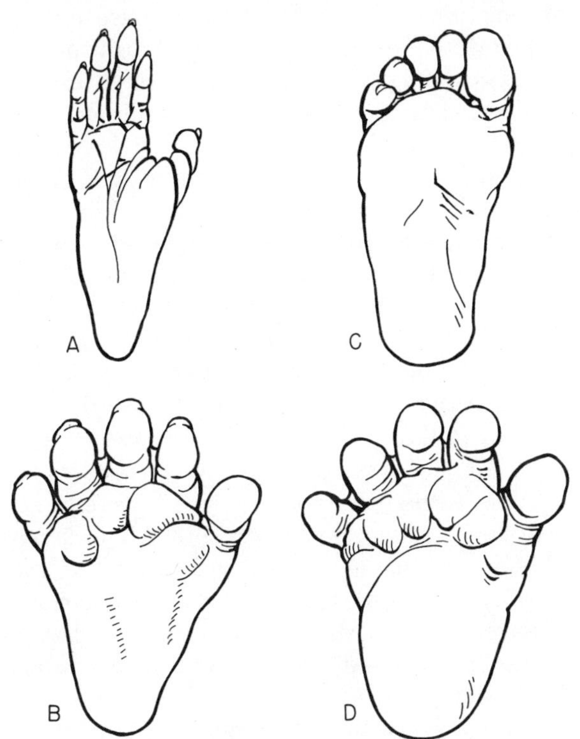

Figure 24. Foot of (A) adult macaque (monkey), (B) early fetus of macaque, (C) adult man, (D) early human fetus. Shows close similarity in fetal stage, but much greater modification during ontogeny in the monkey than in man. (Redrawn from de Beer *Embryos and Ancestors,* 3rd Ed. Clarendon Press, Oxford.)

stead of diverging from that, as it does in other Primates. The presence of numerous sutures between cranial bones after more than twenty years of development contrasts with their much earlier closure in other Primates. The position of the foramen magnum and the correlated cranial flexure is as in the embryo of mammals (including man) but not as in the adults of any mammal except man. Embryos of mammals have a relatively large brain, but in no species except a few small mammals does the adult brain retain as great a size in proportion to the body. Regardless of the adaptive or mechanical reasons for doing so, the failure to develop large eyebrow ridges in the adult skull is simply retention of a feature of the face seen in the late embryo . . . and so on.

These characters, either separately or, more likely, in adaptive correlation with one another, have evolved under the control of natural selection in response to the primary adaptive pressures in human evolution, that is, toward upright posture and prolongation of developmental stages. The means for accomplishing these adaptive results has evidently been to retain certain immature traits because of the selective advantages that they possessed, or they were brought along inevitably as consequences of those that had selective advantages. Several of the features listed are, because of their retarded development, vestigial as compared with their condition in adults of more primitive ancestors.

The rate of growth of different parts of the body is often subject to independent variation, but it is also in many cases allometric; that is, the rate for one part may be different from, but dependent on, that of another (Huxley, 1932). The nose horns of titanotheres (an early, extinct branch of perissodactyls) were thus linked to the total size of the animal. There were no horns in the small, early members of the family, but during their rapid evolution to a size larger than that of modern rhinoceroses, the paired bony horns evolved at a rate even more rapid than that of the body as a whole, so that in the largest titanotheres these organs were of relatively great size. At the same time, the initiation of these structures in ontogeny took place earlier, until their rudiments were present before they could have had any function, in the newborn. The relationship between the growth of horns and that of the body was, in a sense, geometric (more appropriately, allometric) and evidently genetically controlled. As selection favored increase of body size, the horn size was drawn along with it at a different (faster) rate, and the same relationship obtained during individual development.

Among other examples of this phenomenon is that of the "sword" (an intromittent organ) of the male swordtail, *Xiphophorus helleri*, a familiar aquarium fish. Very large males have this organ almost as long as the body, but in diminutive males it is less than half the body length, and in the smallest it does not extend beyond the limit of the caudal fin. Selection for decreased body size can therefore cause this "sword" to become vestigial (or, in the opposite direction, to enlarge). Perhaps a similar cause lies behind the reduction of parts in other dwarfed animals, including the sunfish *Elassoma,* already described. The dwarf salamanders, *Manculus,* of the southeastern states, differ from their closest relative, *Eurycea bislineata cirrigera,* the southeastern two-lined salamander, in reduced size and in the failure of the hind foot to develop its fifth toe, but in very little else (Fig. 25).

Enough has been said to indicate that important structural changes can be brought about in evolution through adaptive modification of developmental rates, and that these vary frequently in the direction of retaining to late ontogenetic stages characters that were seen in earlier stages of ancestors. Inasmuch as this kind of change may take place rather quickly and may not be likely to leave any fossil record, DeBeer (1958) has suggested that the origins of some of the major taxonomic groups of animals have been hidden (or "clandestine") because they were brought about by neoteny. The important point is "the appearance of evolutionary novelties that led their possessors to spread into new ecological zones," by conversion of certain embryonic or larval characters into something altogether different. Examples of this include the change of the uppermost bone of the hyoid arch in a lobefin fish to a small sound-transmitting bone, the stapes, in amphibians, or of the posterior jawbones of reptiles to ear ossicles in early mammals, or perhaps the many-segmented, many-legged body of an early millipede to a short body in the origin of insects (Fig. 26), and possibly the origin of jawless fishes from whatever their protochordate source may have been. (See also Hardy, 1954, on the adaptive significance of neoteny.)

There are some examples of evolution which adds to the ontogenetic sequence of events, increasing the number of steps in development or producing advanced characters in the latest stages or increasing the size beyond that of the ancestor. Formation of bony dermal armor in many of the

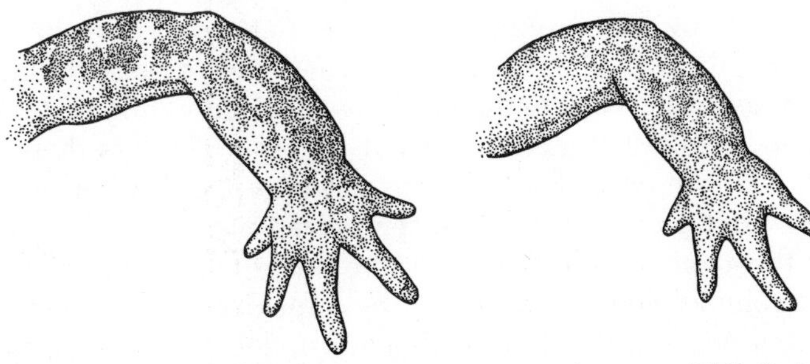

Figure 25. Left hind leg of two closely related salamanders showing loss of fifth toe and reduction of size in the dwarf salamander, *Manculus.*

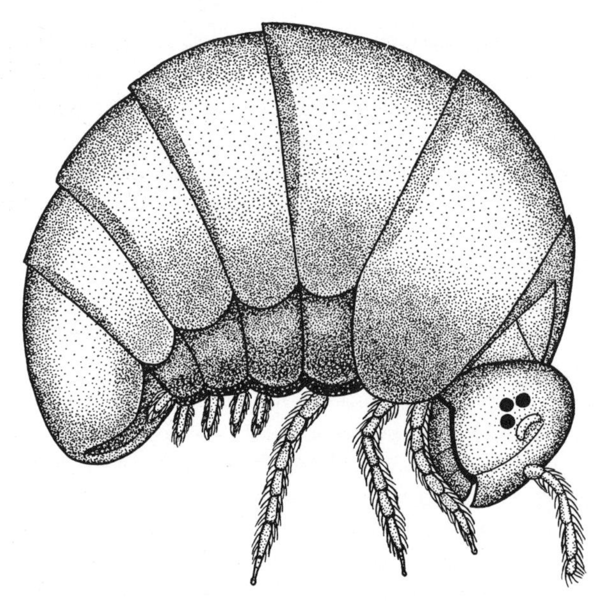

Figure 26. Six-legged larval stage of a millipede.

catfishes of South America (a group in which scales have been lost and the skin is usually naked) is such a case. Sewertzof (1931) cited the development of the beak in the marine, predatory needlefishes, Belonidae; these have at first a greatly elongated lower jaw, as in the closely related halfbeaks, and then the upper jaw, consisting of the united premaxillary bones, grows out to the same length (Fig. 27). I think that Sewertzof was correct in this interpretation, and that needlefishes were derived from halfbeaks, but DeBeer (1958, p. 84) supposes that instead the halfbeaks were neotenic descendants of needlefishes that failed to develop the upper beak. Nicholls and Breder accepted the same conclusion, although they also pointed out that the true flying fishes (Exocoetidae) were derived from halfbeaks by shortening the lower jaw. In the order to which all these fishes belong there is, in halfbeaks and flying fishes, a normal, slightly protractile premaxillary bone in the upper jaw, corresponding in detail to that of silversides, killifishes, and others not far removed. It is therefore most probable that the needlefishes, having a long upper beak consisting only of rigidly fixed premaxillaries, are specialized members of the order. To suppose them ancestral to halfbeaks would be to make, out of an unlikely source, a complex jaw apparatus that by coincidence matches that already present in a primitive halfbeak such as *Chriodorus*. Thus it seems reasonable to suggest that Belonidae diverged in one direction from the halfbeaks and Exocoetidae in another.

In this chapter it is shown that one of the most important categories of genetically based characters is that which affects the rate of development.

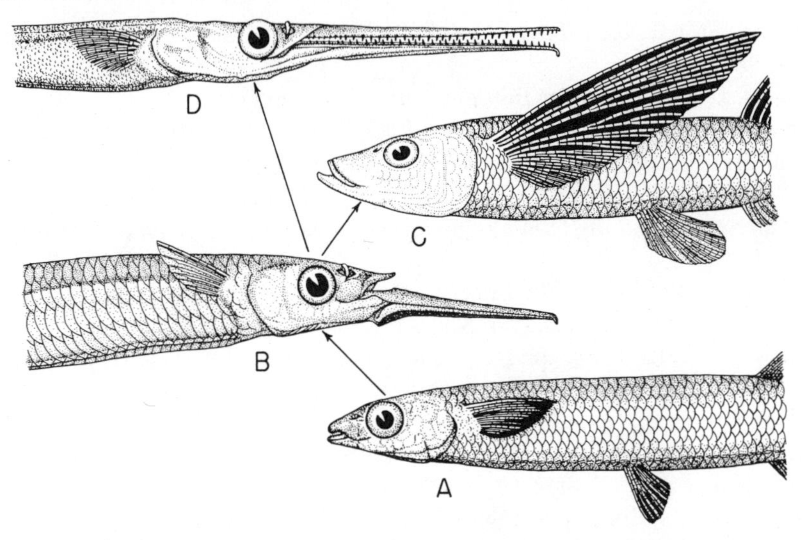

Figure 27. Probable relationships among synentognath fishes: (A) primitive "halfbeak," *Chriodorus,* in which a beak has not developed; (B) typical halfbeak, *Hemirhamphus,* in which the lower jaw is lengthened; (C) flying fish, *Fodiator,* which when immature is like a halfbeak; (D) needlefish, *Strongylura,* in which, after a halfbeak stage, the upper jaw also becomes elongate; this is a specialized rather than a primitive condition, as shown by structure of the jaw as well as other characters of the needlefishes.

Differences in rates can be seen in a few characters, or in many, among individuals, species, genera, and higher taxa. These differences in developmental rates can bring about drastic changes in both the sequence and the nature of characters. The origin of some important characters of higher taxonomic groups can be ascribed to such changes and these, like others, are based on mutations of genes, are subject to selection within populations, and wherever established are adaptive.

REFERENCES

Baer, K. E. von, 1828. *Über Entwicklungsgeschichte der Thiere. Beobachtung und Reflexion.* Bornträger, Königsberg.

Bolk, L. 1926. *Das Problem der Menschwerdung.* G. Fischer, Jena.

DeBeer, G. R. 1958. *Embryos and Ancestors.* 3rd ed. Oxford Univ. Press, London.

Dougherty, E. C., Z. N. Brown, E. D. Hanson and W. D. Hartman (eds.). 1963. *The Lower Metazoa. Comparative Biology and Phylogeny.* Univ. California Press, Berkeley.

Goldschmidt R. 1927. *Physiologische Theorie der Vererbung*. J. Springer, Berlin.

———. 1934. Lymantria. *Bibliographia Genetica* 11:4.

Haeckel, E. 1866. *Generelle Morphologie der Organismen*. G. Reimer, Berlin.

———. 1874. Der Gastraea-Theorie, die phylogenetische Classification des Thierreichs und die Homologie der Keimblätter. *Jenaische Z. Naturwiss.* *8*: 1.

Haldane, J. B. S. 1932. The time of action of genes and its bearing on some evolutionary problems. *Am. Naturalist 66*: 5.

Hardy, A. C. 1954. Escape from specialization, pp. 146–171. *In* J. Huxley, A. C. Hardy, and E. B. Ford (eds.), *Evolution as a Process*. G. Allen, London.

Huxley, J. S. 1923. Time relations in amphibian metamorphosis. *Sci.Progr.* *17*: 606.

———. 1932. Problems of Relative Growth. Methuen, London.

MacBride, E. W. 1914. *Text-book of Embryology. Vol. I. Invertebrata.* Macmillan, London.

Sewertzof, A. N. 1931. *Morphologische Gesetzmässigkeiten der Evolution.* G. Fischer, Jena.

10 · BEHAVIOR AND ADAPTATION

Behavior is the outwardly manifest action of organisms, or of their parts. It varies in complexity from the twitch of a protozoan flagellum to the courtship of whales, the uproarious chorus of a clan of howler monkeys, or the hunt-and-peck operation of a typewriter by a biologist. It is a sequence of events rather than an object, yet it changes during individual development and during evolution in a way that is adaptive, following the same principles of inheritance, variation, and selection as do the other characteristics of organisms and having to about the same extent a genetic basis.

Study of animal behavior has passed far beyond the folklore and anthropomorphism in which, up to the end of the nineteenth century, most people attributed human ideas and motives to animals. It is important in discussing behavior that we make no assumptions of purpose, mind, or even consciousness unless the evidence shows that such concepts are required. Behavior is a part of physiology. It is the action of the whole or of some portions of the animal; it cannot be separated from anatomy or from development; it cannot be treated abstractly, as if it existed alone.

Behavior is usually a part of an adaptation and is better understood thus than when taken out of context. For instance, a particular unit of behavior, which we might call an act, may be performed by an insect in a characteristic sequence or pattern as a result of a stimulus, such as food. Like many adaptations of structure or physiology, this behavior is genetically controlled, innate, with little variation. Its slight differences among individuals of a species or among species of a genus may be compared in the same way that structures are compared, making a record or description in each case. From these, conclusions may be drawn as to adaptive meaning or, sometimes, evolutionary relationships.

On the other hand, in many animals the response to a stimulus may be any of a number of different actions, more or less appropriate to the circumstances but not wholly predictable or patterned. If certain actions result in discomfort or frustration, the same individual meeting this stimulus on future occasions may inhibit the inappropriate responses and come to act in a manner that appears patterned and leads to a satisfactory result. The ability to learn, to establish a habit, to settle upon one among many possible alternatives as the adaptive one is itself an adaptation and is inherited, as are other adaptations. The process of learning in an individual, by inhibiting unsatisfactory responses and substituting a correct response, appears to belong in the same category as physiological adaptation, already discussed. Although not comparable in detail to many examples of this, nevertheless learned behavior operating on choice of alternatives allows an individual to adjust itself to a specific set of conditions.

Not only does behavior evolve, but it takes an active part in ecological differentiation, divergence of species, competition, selection, replacement, and so on. It is subject to adaptive radiation within a group that has a common ancestry, to convergent or parallel evolution, to changes in a linear fashion, to increasing complexity or simplification in both evolution and ontogeny. Study of its evolution has to be based mainly on observation of animals now living, subject to control by information on their actual relationships obtained in other ways. This does not make evolutionary judgments impossible or even "speculative," for there are ways of determining which patterns are primitive, which specialized, and of correlating these with parallel judgments on morphological or other features. A great deal can be inferred about some aspects of the behavior of various fossil groups whose evolution and relationships are understood. But for very few groups either fossil or living is there thorough, detailed, descriptive information on behavior. This is our foremost need.

The current surge of interest in this field has brought a diversity of people together. All have contributions to make, but they differ in their acquaintance with animals, their techniques, preconceptions, and habits of thought. We may expect disagreement far into the future; the various bases of behavior and the full course of its evolution will not be fully known for many years. Students, to have a fair perspective, should be conversant with other aspects of the subject as well as those that concern them most directly. No doubt we shall arrive at some agreement, explain some parts of behavioral evolution, without knowledge of all the factors. Greatly needed are close comparative studies among related species, frequent reviews of progress, and critical summaries.

SOME TERMS AND IDEAS

Ethology. The study of behavior is expanding rapidly and the language is changing. A substantial body of knowledge has accumulated in the last 40 years on the behavior of particular species of animals in natural, or approximately natural, conditions. This is called ethology. It arose largely from the work of Lorenz and Tinbergen and their followers in Europe and America. It differs from the old-fashioned folklore of behavior, and from the experimental, laboratory type of study in which the kind of animal is less important to the experimenter than the purpose and conditions of his experiment. Ethology uses prolonged, precise, repeatable observation under conditions that are controlled (when possible), and the data are objective. Ethology has produced its own working hypotheses and special terminology, some of which may need revision as time goes on. The results of ethology so far are most adequate for a few kinds of insects, a few fishes, and certain birds and mammals that can be watched for long periods. But as yet the quantity of information needed for a broad picture of behavioral evolution is not available for most groups.

Innate Behavior. It is convenient to designate most of the behavior of most kinds of animals as innate, meaning that the animal's response to a particular stimulus under given conditions is fixed by its heredity and does not vary appreciably. This includes the simplest elements of behavior (for example, ciliary action) as well as some of the most complex patterns (production of a geometrically precise web by an orb-weaving spider). Subjecting animals to diverse stimuli or to abnormal environmental conditions can, of course, alter the manner in which the actions are performed, or prevent them, but such alterations are not learning, and the form or pattern is not learned. (A precautionary note: When a biologist speaks of studying the "habits" of birds or other organisms in natural conditions he is understood to be using a short, common, and legitimate expression about which no quibbling is necessary. He need not explain to other biologists that he means behavior in general and is not talking about any kind of learned behavior, such as the *habit* of smoking cigarettes.)

As we examine the behavior of "higher" animals, especially birds and mammals, we find it is not so completely innate or fixed as in simpler kinds. This is notably true in the larger-brained placental mammals, in which an ability to make varied responses to stimuli becomes obvious, and the possible alternatives in doing so may be numerous. There remains, especially in young stages, a controlling innate simplicity, upon which learned responses may presently be built.

Taxes and Tropisms. Tropisms are simple, directional movements either toward or away from the stimulus that induces them. The term is generally limited to movements of growing plants: toward a source of light in the case of green leaves, away from the pull of gravity in young seedlings, toward gravity in the early stages of root development, and so on. It has often been applied to animals that show directed movements, but as the latter involve locomotion of some complexity they are now usually distinguished as taxes (singular, taxis). The location of the source of stimulus determines the direction of movement, which is either toward (positive) or away (negative). The commonest kind of directional stimulus is probably light, and it is a familiar experience that moths and other night-flying insects turn and approach an artificial light, sometimes going by, swinging around, and coming irregularly closer until they flutter against it. But the same moths do not respond in that way to light from the sun during the day; indeed, for most insects nothing comparable to an artificial light ever normally comes into their lives. Thus such behavior can hardly be seen as adaptive. But for many phototactic animals there is adaptive meaning: Flies resting in a dark place may become positively phototactic and emerge into the light as the temperature rises, and cooling produces negative phototaxis; a large proportion of the swarming planktonic life of the sea is negatively phototactic, sinking deeper in the water by day but rising to the surface after dark. Many other examples of taxes are known, especially among invertebrates.

Reflexes. In its simplest meaning the word "reflex" implies that a stimulus affecting some part of an animal causes impulses to go in along nerve fibers and, so to speak, be "reflected" outward to an effector organ which is activated by the impulse either to move or to secrete. The pathway of impulses in most animals is from their origin in a receptor (that which receives the stimulus) through the afferent (sensory) and the efferent (motor) neurons, including en route some part of the central nervous system, and terminating in muscle or gland cells that respond.

Evidently actions quite comparable to this occur even in Protozoa and may be considered fundamentally similar even though there are no neurons. For example, *Paramecium* coming in contact with an object while swimming stops, reverses the action of its cilia, backs away, stops again, and goes forward in a different direction. This may or may not cause it to avoid striking the object again.

Stimuli that induce reflexes are either external or internal and consist of some kind of change in the energy that impinges on sensory receptors or neuron endings—usually a change of intensity. In the long run, "stimulus" can hardly be defined except by its result; it is *that which causes an impulse to start.* It can be any of a great variety of kinds of energy, but by no means all, and some animals have receptors that others lack. Color vision is due to the differing photochemical effects of light of various wavelengths on the visual pigment in cone cells of the retina. Primates, like birds, have highly developed color discrimination, but there is little or none in other orders of mammals. Red light is able to initiate impulses that are not initiated by blue or green or gray, but in animals without color vision wavelength has no effect, and any light evidently initiates the same nerve impulses.

The impulse itself does *not* differ according to the kind of stimulus or the intensity; it is subject to the "all-or-none" law, and consists of a wave of negativity in the ionic charges of the cell membrane, proceeding rapidly along a neuron fiber at a rate determined (roughly) by temperature but not as fast as an electric current. (In mammalian neurons it is about 100 meters per second.) An impulse does, of course, cross the synapses between two or more successive neurons. This is done, in vertebrates, by secretion of acetylcholine (or of sympathin) at the farther ends of neurons. On arrival at the effector cell, its response is produced in the same way.

Instinct. Although this term is of little or no use in psychology, it may have some validity for animal behavior if its meaning is carefully restricted. Instinct does not mean a mysterious kind of knowledge, it is not a power to foresee the future, it does not tell birds to go south in the fall because winter is coming, it is not a drive or an urge to accomplish a particular purpose; none of these older meanings has any basis in scientific evidence. To substitute the word "drive" does not make such concepts more valid. But we may properly use "instinct" to designate (not to explain) a pattern of behavior that is recognizable as an adaptive unit, characteristic of a species under specified conditions.

The difference between an instinct and a series of simple reflexes can easily be demonstrated by the defensive behavior of a crayfish in a confined space. When it is touched, if it has no room to retreat, it rears back on its legs, raises its two big claws and opens them, attempting to pinch any object that is close enough. Evidently the senses of touch and sight, at least, are involved in this coordinated pattern of behavior. A crayfish, like other arthropods, has a central nervous system consisting of ganglia connected by paired cords along the ventral side of the body, one pair of ganglia per segment, and from these extend nerves to the muscles of the legs and other parts of each segment. Any activity involving more than one segment might be expected to employ impulses that pass from ganglion to ganglion through the ventral nerves. If these nerves are severed between successive segments, the ability of each segment to act is not impaired, provided the stimulus is given to the segment or an appendage of it. If a leg is stimulated it is withdrawn, and reflex action of one or both legs may occur. But no stimulus applied to the body will induce any response in parts that do not belong to the same segment. No coordination of segments is possible. The reflexes are not lost, but the defensive behavior just described is impossible because it requires coordination of actions of most, or all, segments and appendages. The instinct is the whole; reflexes are its parts.

Among countless other examples of instinctive behavior is the complicated courtship of the balloon flies (Empididae), so called because the males of some species make a balloon-like bubble which they carry to present to the female as an incentive to mating (Kessel, 1955). The evolution of this behavior can be traced in a series of species back to the original capture of a small insect, which is offered to the female. Apparently there has been selective value in the various refinements by which the gift was gradually transformed into an artificial and useless object. Clearly also, this, like other mating instincts, is reciprocal between male and female, and its evolution in the two sexes must have been a sequence of changing but coordinated responses of one to the other. Indeed, the full pattern of this instinct is not manifested by one individual but by a pair, and is not complete until the two have separated and have no further contact. A new sequence then follows in the female,—that of egg laying.

The probable evolutionary sequence of complex instinctive behavior among related species may in many cases be inferred with considerable assurance even though the species are, necessarily, contemporary. In any large, diverse group of animals there are some species that retain characters of behavior (as of structure) that are little changed from those that must have been ancestral, and whenever the whole pattern has a sufficient complexity to be viewed as a configuration of many details, then the way in which changes took place in it may become evident in making comparisons.

Learning. The "learning" of invertebrate animals, as far as it has been described, is by "inhibition of responses and response tendencies which block the animal or fail to lead it to some terminal response, such as eating or escape from noxious stimulation" (Harlow, 1958). That the behavior of invertebrates is in some cases not altogether fixed but can be modified as a result of experience was discovered by patient and carefully planned experiments rather than by casual observation. For example, in a marine flatworm, *Leptoplana,* phototaxis was inhibited by a touch stimulus given whenever the worm, stimulated by a light, started to creep in response to it. The interfering stimulus was not enough to injure or fatigue the animal but only to cause it to stop moving, until, after many trials, the worm learned not to make the normal phototactic response. In a similar way, the normal response of tentacles of a sea anemone to touch was inhibited after bits of filter paper were used, repeatedly, for the stimulus. Normally the response of tentacles to any contact is to bend toward the point stimulated so that, for example, an item of food can be captured. This happens in the immediate vicinity of the contact and not over the entire disk of tentacles. When the stimulus is not food but bits of filter paper applied again and again to a certain group of tentacles, these but not others learn to inhibit their response to this particular stimulus. The learning persisted in these tentacles for a period of 6 to 10 days, after a few trials. Another example given by Harlow is the inhibition of the response of an octopus to bait attached to an electrified plate, a slight shock accompanying each response, in the course of 25 trials; at the same time its response to bait not electrified was uninhibited.

In these examples the responses that were attended by unpleasant interference, by uncomfortable stimulation, or by failure to get a satisfactory result were gradually eliminated. On the basis of this kind of experimentation Harlow infers that learning in general is in some way inhibitory; the organism has the potentiality of responding in one way, or another way, perhaps various other ways, or not at all, to a given stimulus, and when the "wrong" responses are "punished" or not rewarded, the result is eventually to suppress all except a response that is not so punished. Certainly this concept has the virtue of simplicity, and with some thought one may find it applicable to many learning situations. But we may require a good deal more information before accepting it as adequate for all learning.

Is it, for example, consistent with or contradictory to the common idea that learning can result from reward as well as from punishment? As far as the experimental animal is concerned, reward and punishment are not necessarily opposites, but might be considered rather as different points on a scale of values that extends all the way from extreme pleasure through minor degrees of satisfaction to indifference, mild discomfort, great distress, and unendurable suffering. Just as heat and cold are relative terms on a scale of temperature, so "punishment" might occupy almost any point below the highest satisfaction on this scale of rewards. For the animal's response to be attended by no consequences at all could, in other words, play the same role as a noxious stimulation, when contrasted with the height of pleasure. Therefore Harlow's theory appears to be entirely consistent with the theory of reward and punishment, or "pleasure–pain," as an explanation of the change of behavior of an animal in learning.

The theory implies that there is always more than one possible response to a stimulus. Either more than one effector can be reached by impulses from one receptor, or they may or may not be blocked en route. Among the simpler invertebrates the number of alternatives is small, seldom more than two or three, but in animals of complex structure and multiple activities the number of possible responses to a given stimulus increases greatly. During evolution in advanced phyla of animals there has been a general increase in refinement of receptors and complexity of nervous connections, along with adaptive diversity of the locomotor and other effector organs. In specific cases, of course, adaptive characters including those of behavior may secondarily be reduced or lost, under circumstances in which doing so is of advantage. Loss of the structures and activities associated with walking took place in several groups of vertebrates that lost their limbs: the snakes, glass lizards, amphisbaenids, and some eel-like salamanders. A loss of eyes and therefore of behavior related to sight accompanies adaptation to cave life in some species of crickets, beetles, a few freshwater fishes, and cave salamanders. But on the whole, evolution has resulted in increasing complexity of characters related to behavior.

Insects and vertebrates are, in quite different ways, the two most progressive, successfully adapted, groups of higher animals. (The fact that insects are a *class* of arthropods, and that vertebrates are a *subphylum* with at least eight classes of chordates, is immaterial to the comparison, as these taxonomic ranks do not correspond in any real way in groups so distantly related. Vertebrates as a whole show less structural adaptive diversity than the insects, as well as a much smaller number of species, say 60,000 against more than a million.) But among insects, although behavior becomes exceedingly complex in a few groups, notably those that are social, its evolution has included very little development of the ability to learn. Insects are not intelligent even if they do certain things that we, as intelligent animals, can learn to do also.

The advanced vertebrates, on the other hand, have in some lines evolved a remarkable complexity of the central nervous system, with a capacity for numerous modifications of behavior. Instead of two or three possible responses to a stimulus the number may be very great, and individual adjustment to circumstances may therefore be brought about in part by learning certain appropriate responses to make, inhibiting the others. This ability, which is an adaptation itself, has evolved independently in quite different ways in several unrelated orders of mammals. Although the ancestral placental mammals of the late Cretaceous and early Cenozoic were intelligent compared with the average reptile, their brains were small and simple relative to those of most modern species. The earliest horses, for example *Hyracotherium,* had a brain closely comparable in size and structure to that of an opossum, *Didelphis.* The same was true with the ancestors of primates, elephants, carnivores, whales, and the various orders of ungulate mammals. Independently in these groups the brain evolved not only a greater size but also complexity of the cortex of the cerebral hemispheres, to a condition of high intelligence in some advanced members of these orders. This does not mean that the intelligence of a whale, a horse, or an elephant can be measured in the same way, compared directly with each other, or expressed by the same acts of learning as that of a raccoon, dog, or man, because these differently built and differently adapted animals cannot be given the same kind of tests. A whale cannot be exposed to the same problems as a cat in a puzzle box.

ADAPTIVE NATURE OF BEHAVIOR; EVOLUTIONARY RELATIONSHIPS

From the beginning of its life any animal is engaged in meeting simultaneously various different aspects of its environment and is equipped in its physiology and structure to sustain a more or less prolonged relationship with them. Much of the environment usually has little or no significance to the animal at any time, but to certain parts of it the animal can and must respond to continue living. Food of one kind or another, for instance, is available but not uniformly distributed, as a rule, throughout the surrounding medium. Particular activities may be required to locate food and get it into the body, to reject what is not usable, and to eject what is not digestible. There may be external forces of gravity, of moving water or air, of impact or pressure, and of contact with other animals or plants, and to these forces animals react in a way that maintains for them a favorable status. The environment may contain dangerous or noxious features: heat, cold, strong light, poisonous substances, deprivation of some essential components of the medium, parasites, and predators. Depending on the frequency of these features of the environment, a part of the behavior of an animal consists of responding to them by avoidance, by approach, by testing or exploration, and by movements that are actually (as well as apparently) random. These, then, are reasons why behavior is adaptive and why it is incorporated in many structural and functional adaptations in animals.

To paraphrase Pittendrigh (1958), insofar as living things are organized they are adapted, and to the extent that behavior is organized it is adapted. This organization is relative to an end, which does not imply teleology but *adaptive meaning*. The meaning, ultimately, is survival and reproductive success. Pittendrigh remarks that the organization of a hen for getting food and avoiding danger is directed toward the supreme end of laying an egg (that is, perpetuation). This expresses the essential meaning of natural selection. It is not a force or a directing agency but an abstraction of these: what survives and perpetuates itself is that which is organized to do so. Survivors are the products of efficient reproducers and are "effective vehicles of their genotypes."

As far as natural selection is concerned, it has produced and maintained adaptation through the available genotypic possibilities in a population that relate to its particular environment. In this sense adaptation *is* the inherited organization, including capacities for behavior, received by individuals in each generation. Thus a specific pattern of fixed behavior, such as that of the balloon flies already mentioned, is an adaptive unit of behavior in exactly the sense that parts of a body organized for a given essential function constitute an adaptive mechanism. Further, the fact that the pattern is inherited as a unit, subject to the normal occurrence of an appropriate stimulus, means that behavior adaptations can be traced among related species from a specialized or advanced type back, in many cases, through several stages of evolution that have been retained among organisms still existing.

The other meaning of adaptation in behavior applies to those kinds of behavior that are modifiable in individuals as a result of experience. Just as the eyes of many mammals have the inherited capacity to respond to varying intensity of light by narrowing or enlarging the pupil, so animals that are capable of learning have the inherited ability to change their response to a particular stimulus if a given response, repeatedly, results in frustration or discomfort, whereas other possible responses to the same stimulus do not. There may be a gradual inhibition of the unsuitable response and replacement of it by an alternative that is more nearly adaptive, more satisfactory to the animal. Thus an individual, by learning, can adapt its behavior to circumstances, just as the iris adapts the eye to receiving a moderate but adequate intensity of light. It seems appropriate to speak of learning as a special kind of physiological adaptability. There is no evidence to support the old idea, current at the beginning of this century, that innate, fixed behavior could originate from learned habits being incorporated, somehow, in the "germplasm," thus becoming "instincts." (See the discussion of the Baldwin effect, p. 113. For another kind of relationship between behavior and adaptation see the discussion of isolating mechanisms, p. 74.)

To the extent that the behavior of a species contains elements that are distinctive and do not occur in closely related species, they can serve as taxonomic characters in the same way as distinctive morphological features, and the same is true for any higher category above species. Examples of this for species are given by Dilger (1960, 1964) in the African parrot genus *Agapornis*. *A. personata* carries nesting material in the beak, but *roseicollis* tucks the material among the feathers of the rump. "Our hybrids between these two species basically inherit the act system of *roseicollis* but certain of the acts are missing or weakened in their performance so that these birds never succeed in carrying nest material in their feathers, even after years of trying. There is a slow increase of carrying in the mouth so that in about four years they are nearly as efficient as *personata*. Experiential influences are responsible for this shift but they are never able to improve the act system of tucking strips amidst the feathers to the point where it become operative" (Dilger, 1964). Dilger's work shows that what he calls "experiential influences" have a part in forming the species-specific behavior of these birds. This means that the characteristic differences among species are not entirely due to the difference in genotypes. From the point of view of taxonomy this would not matter, provided the elements of experience were as normal and characteristic as the genotypes themselves.

The evidence is almost never sufficient to indicate how much of the nongenetic component of behavior is actually learned. If all the stimuli that have ever affected the organism in its life constitute "experience," then it would be quite true to say that there is no behavior without prior experience, and this could be stretched to mean that there is no innate behavior at all. But it cannot be shown that most of this experience has any effect on subsequent behavior. Only that part is relevant which causes the reflexive or unlearned components to be changed or replaced by others. The problem becomes difficult to handle here, but the term "experiential influences" does not seem to have enough meaning to be used as a substitute for "learning." It implies, but does not demonstrate, that *any* experience is relevant to later behavior.

In the course of adaptive radiation of "Darwin's finches" in the Galápagos Islands, the form of beak is correlated with feeding behavior. Major differences of color, pattern, and form among Lepidoptera are usually associated with corresponding differences in behavior: In the large family of geometrid moths, most of which fly by night, those that have a gray, brown, streaked, or mottled pattern are the ones that spend the day motionless on the trunks of trees or among dead leaves on the ground, visible but not noticed because of resemblance to their background; the few pale green species also fly at night but are seldom seen by day because they rest among green leaves. If these insects were active by day and came to flowers, walked about on the ground or on tree trunks, or rested at random in places not similar in appearance to their own patterns and colors, they would be eliminated rapidly by predators. To remain motionless on a particular background is an essential element of their adaptive behavior. By a parallel behavioral evolution, this habit is adopted independently by numerous other families of moths, by many beetles, bugs, and spiders, and for that matter by caterpillars (regardless of the habits of the adults).

The behavior of a species, a genus, a family, or even a higher taxon may have some features that constitute reliable characters for recognition of the group, but this reliability is often limited by the occurrence of the same behavior in other groups, independently. Therefore such characters are most useful for evolutionary or systematic understanding to students who have wide familiarity, in the field, with the groups in question and who have learned to be cautious in judgment.

A good example of the use of a behavioral character to revise a systematic arrangement previously based on morphology is cited by Mayr (1958, p. 345). The crag martins, *Ptyonoprogne,* should be placed near *Hirundo* (barn swallows) because they build nests of mud and agree in voice and the presence of concealed white tail spots; they were earlier placed with *Riparia,* the bank swallows, which dig tunnels in banks. This does not mean that a behavioral character deserves greater weight than a structural but that it can be a signpost to suggest conclusions not clearly indicated by previous work.

Geographic variation of behavior within species has not been investigated widely as yet. One instance occurs in the polyphemus moth, *Antheraea polyphemus*; its caterpillars in the southern states attach their cocoons to twigs of trees by a broad strap, but in the north they are not attached and drop to the ground. Voice differences within species are common; for instance, the hoot of the barred owl has four syllables in the north, but in the middle and southern states there are five ("who cooks for you-all?"). These characters seem to be as distinct and recognizable as the common features of color or form among subspecies.

COMPARATIVE STUDIES OF BEHAVIOR

The evolution of special kinds of behavior has been studied in a few cases in recent years, and the number will probably increase rapidly, but such investigations require prolonged and elaborate effort. One has been mentioned already, the courtship of balloon flies (Kessel, 1955). A symposium on the evolution of construction by animals (*American Zoologist,* May 1964) presents a number of examples in which the morphology of nests, burrows, cases, or other structures serves as a record of complex behavior in the first place, and in the second as a means for detailed comparison of evolutionary stages in the history of such behavior. To the objection by some biologists that the past is unknowable, that we cannot infer early evolutionary stages of a characteristic when our evidence is wholly contemporary, the answer is that the contemporary evidence itself indicates some characters to be more primitive than others, that we can frequently (not always) choose the more primitive among two or three different alternative states, and that such choices made among several sets of characters can often be shown to be mutually consistent rather than contradictory, thus making a given evolutionary interpretation highly probable rather than "merely speculative."

Evans (1962) investigated the evolution of prey-carrying mechanisms in wasps, showing that the more advanced methods of transporting prey from the place where it is caught to the nest where it will be stored result in less delay, less chance of loss or attack on the way, and thereby increase the range within which hunting may be done. The trend in several different families of wasps has been to increase the number of prey provided in the nest for each larva of the wasp.

The primitive stage of this behavior in spider wasps is to drag the prey backward toward the nest, but some spider wasps, without special mechanical adaptation, also carry a spider in the mandibles, part way, while flying. *Anoplius depressipes* flies, dragging the large water spider *Dolomedes* on the surface of still water. *Priononyx,* a primitive digger wasp (Sphecidae), drags short-horned grasshoppers to the nest by their antennae, walking forward and straddling the grasshopper; one is provided per cell. *Sphex* brings two or more long-horned grasshoppers per cell, carrying them in flight; each one is held by the mandibles and all the legs. *Philanthus* generally uses either the mandibles and middle legs or the middle legs alone, carrying prey, and provides several insects per cell. The cicada killer, *Sphecius speciosus,* may start transporting its very large prey by using the middle legs, flying from a tree but coming to the ground and then carrying the cicada the rest of the way by walking. Tibial spurs of the middle and hind legs help support the cicada. In *Oxybelus* the prey is carried in flight by the hind legs of the wasp and held far back; in some species it is impaled on the sting as well. Certain other wasps, *Aphilanthops,* have a modification of the posterior abdominal segment for holding the prey; the sting of the wasp is also inserted between two pairs of coxae of the prey, holding it out behind the wasp so it will not interfere with flying.

Several different phyletic lines of wasps have separately evolved through a series of stages in the manner of carrying prey, beginning with the primitive habit of dragging it backward holding by the mandibles; then walking forward carrying by mandibles, with flight in some cases; flying with prey held in mandibles, with or without the sting; dependence on the sting with some support from the hind legs; sting alone with or without abdominal clamp. Several lines have gone through the first three stages, four have reached the fourth stage, and only two have gone beyond the fifth, according to Evans.

Among recent studies of nest construction by animals is one by Schmidt (1964) on the African subterranean termites *Apicotermes.*

> In contrast to the constructions of some animals, the nests of *Apicotermes* are remarkably species-specific. . . Each species builds a very characteristic type of nest that is distinct from the nests of even closely related species. The nests are complex, thereby providing an unusually large number of characteristics. . . In fact, the nests of *Apicotermes* provide much better material for phylogenetic studies of this genus than do the termites themselves. . . The fact that termite nest-building behavior is innate is shown by the method of colony founding. A new colony is started by two termites, a king and a queen, that pair and then seal themselves within a small cavity. The queen eventually lays eggs that produce workers. These workers then build a species-specific nest, although they have never experienced such a nest or previous generations of workers.

In a similar vein, Michener (1964), discussing evolution of the nests of bees (Fig. 28), concludes: "Comparative studies of nests show not only distinctive features of the taxa from subspecies to superfamily, but also homologies among nest structures. Such homologies can be supported in detail for related forms by ontogeny, behavior patterns involved in construction, as well as relative position, appearance, and function of nest parts but suffer from the same problems of definition as homologies of structural parts of organisms themselves."

DEVELOPMENT OF BEHAVIOR

As would be expected from much of the preceding, we cannot separate in any clear-cut way the development of behavior from that of morphology. During early embryonic stages the morphogenesis of receptors, of the nervous system, and of effector systems proceeds in such a way that the earliest activities take place when the development of appropriate parts has reached, synchronously, the functional stage. This was shown in an elegant manner by Coghill (1929) in the embryo and early larva of the salamander *Ambystoma tigrinum*. The first movement is a slow bending of the head to one side and then the other by contraction of the anterior muscle segments of the right or left immediately behind the head. It is about 36 hours before the segments of the entire trunk and tail reach an active stage and are able to bend the animal into a tight coil. The speed of action also increases in this time. After this the lateral bend in one direction at the anterior end is followed immediately by a bend in the opposite direction before the first bend has passed through the entire length of the animal. Instead of a coil this produces a double curve, S-shaped, each part of which progresses posteriorly until it disappears at the tail end.

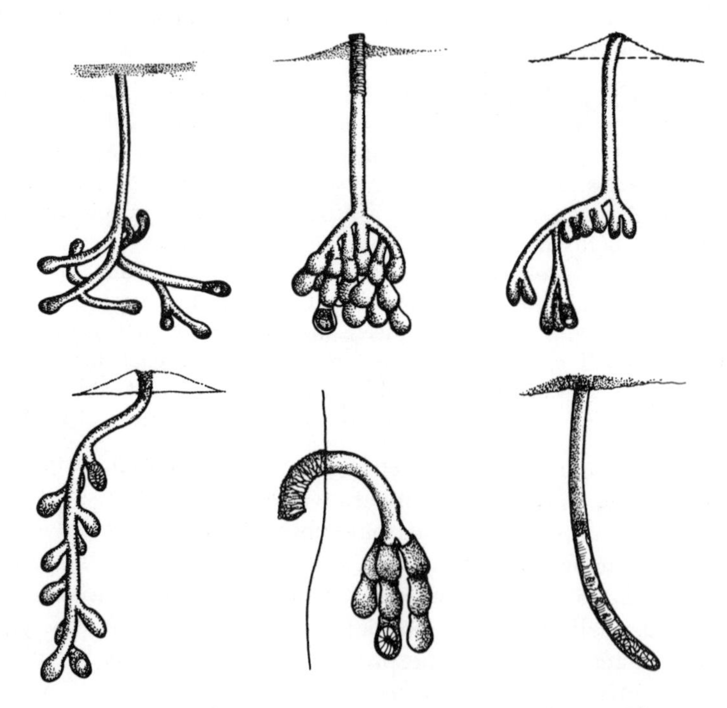

Figure 28. Nests (burrows) of solitary or primitively social bees.

At about this time the salamander hatches from the jelly capsule in which it has lain and becomes a swimming larva. Posterior movement of the waves of contraction causes corresponding pressure against the water and the animal swims forward as a result. Development of this sequence of behavior of parts of the body results, first of all, from the arrival of the tips of developing motor neurons into contact with newly formed muscle fibers of the most anterior segments. Meanwhile sensory neurons grow out from the dorsal longitudinal tract in the spinal cord and soon are able to transmit impulses from touch stimulation of the skin and stretch stimulation of the muscle segments. By ventral cross connections in the cord these impulses are transmitted to muscles on the opposite side of the body, either initiating a primary flexion or inducing the alternating flexions of one side after the other. At this time the crossing over of impulses from the sensory system of one side to the motor tract of the other side takes place in the anterior part of the spinal cord or posterior part of the brain, and the resulting action moves, wavelike, from the anterior muscle segments down the body toward the tail.

This is not the whole story of the early development of swimming behavior in *Ambystoma,* but it is enough to show that behavior is organized in relation to a sequence of stages in the growth of the muscular and nervous systems. It is thought that the terminal and synaptic connections of developing neurons are established in a quite specific pattern from the beginning. When developing or adult nervous systems of experimental animals are disconnected surgically in various ways, the result is specific malfunction or loss of function that is not corrected by later experience. Many neurologists believe that to some extent in young children different neuron pathways may be substituted (or associations learned) that can compensate for partial brain damage, but the way in which this is done is not known. In older children and adults such damage can be corrected only by learning to use substituted mechanisms, that is, different muscles and movements.

In the animal kingdom as a whole the development of behavior in individuals may either follow a fairly direct sequence, if such is the sequence of morphological development, or may be greatly modified at different times in the life cycle if there are larval stages adapted to different ways of life. For example, in the metamorphosis of holometabolous insects there is an interruption of all outward activity, a quiescent period, and a complete breakdown of virtually all internal structures during the pupa stage, after which the general morphology, physiology, and behavior are completely reorganized for the very different life of the adult. Adaptations of the animal are organized once for the young stages and then all over again for the adult.

GENETICS OF BEHAVIOR

The degree of genetic control of various types of behavior is extremely varied, often precise, but in other cases affording much flexibility. It is proper to speak of inheritance of behavior just as we refer to inheritance of red eyes, normal wings, color of flowers, shape of seeds, or any phenotype for which a genetic basis is known. Biologists understand that flowers, wings, or reflex actions are not transmitted as such from parents to offspring, but they do not quibble about using the word "inherited." To a small or large extent all these characters are subject to modification during development because of environmental circumstances, and of course any pattern of behavior must await the completion of processes of development that make it possible. Finally, it is only to the extent that behavior has a genetic basis, and is therefore affected by natural selection, that it evolves and is adapted.

The study of this genetic basis is very recent, few papers having been published before 1950; the first book was that of Fuller and Thompson (1960). In the *American Zoologist,* May 1964, there is a series of papers constituting a valuable "refresher course."

Bastock (1956) was able to explain an effect of the gene for yellow body in *Drosophila* noticed long ago (Sturtevant, 1915) that reduces mating success of the males. The duration and strength of vibration of the wings in courtship is diminished in males by the gene *y,* and as this vibration, transmitted by contact of the antennae, is normally a stimulus to the female, the frequency of matings is lowered. Scott (1943) found that most single genes (those that, acting alone, produce a particular phenotypic effect) in *Drosophila* also have some influence on behavior. It is now thought that the vast majority of genes have more than one effect; this is pleiotropism. Any disturbance of sensory, nervous, or motor structures or functions might readily modify behavior. In some cases it may be possible to recognize a prior cause of some phenotypic character, and presumably all pleiotropic effects of a gene have their origin, directly or not, in a particular biochemical action early in the development of the individual. But we seldom have enough information to determine what this primary process is or to trace the steps by which it produces the various pleiotropic characters. Ordinarily, then, we cannot say that one character is the "primary effect" of the gene and others are pleiotropic, in the sense that they are secondary to it. Preferably, all the effects have a pleiotropic relationship among themselves, and we may avoid the word "primary" until it is demonstrable.

In the flour moth, *Ephestia* (Caspari, 1958), there is a pair of alleles that has both morphological and behavioral effects. *Rt* causes brown testis and higher mating frequency in the males; *rt* produces high viability and, when homozygous, red testis. Therefore, males with *Rtrt* have brown testis, high mating frequency, and high viability; this combination of mating frequency and viability in the heterozygote is advantageous, so both alleles are maintained in the population, an example of stable polymorphism.

An ingenious series of experiments by Rothenbuhler (1964) demonstrates the simple mode of inheritance of "hygienic behavior" of worker bees and shows that another kind of behavior, stinging, is inherited differently. Of course, among honeybees, unlike almost all other animals, less than 1 percent of the individuals in a population can reproduce. A hive contains primarily sterile female workers which are the offspring of a queen and drone. None of the characteristics of workers are inherited directly from workers, nor can any of them have progeny. There is variation in behavior among workers in a hive, but there is more variation between those of one hive and those of another, and there may be quite distinct differences in behavior between various races of honeybees.

The experiments by Rothenbuhler concerned the behavior of two different strains of bees in cleaning the hives; this consisted of removing dead larvae, after uncapping their cells, when the larvae had been killed by the infectious disease called American foulbrood. In the Brown Resistant Line, there was complete removal of the dead larvae, and in the Van Scoy Susceptible Line only a few were removed, nearly all being left in their cells. The technique was this: "A sample of eggs from a suitable mated queen is obtained in a comb, and is placed for hatching and rearing in the colony whose behavior is to be studied. At the appropriate larval age the larval food surrounding each larva is inoculated by means of a microsyringe with spores of the pathogen dispersed in water. Control or check larvae are inoculated with water only or are uninoculated as the situation requires." Several hundred larvae were in each of these categories. In three colonies of Brown Resistant the behavior was as indicated above, and in four colonies of Van Scoy Susceptible it was nonhygienic, as indicated. Then similar tests were made in five colonies that were the F_1 offspring of crosses between these two strains. As a result, the behavior was in each case nonhygienic, as in the Van Scoy line, suggesting that this behavior is genetically dominant and the hygienic is recessive. But backcrossing was necessary to be sure. The procedure and results are of great interest:

> F_1 queens were necessary for reproduction. From such F_1 queens drones were obtained. These drones, having developed from unfertilized eggs, were gametes, gametes with wings, in fact, and the capability of producing 6 to 10 million more genetically identical gametes in the form of sperms. Such a gamete from an F_1 individual, when mated by artificial insemination to an inbred queen of the parental line, will produce a colony of bees of similar genotype. 29 such drones or gametes were mated to 29

> queens of the Brown Line, which were expected to be homozygous for the recessive genes for hygienic behavior. If ½ of the 29 colonies developed from the backcrosses were hygienic, one would conclude that one locus only was involved in the difference in hygienic behavior between the two lines. If only ¼ were hygienic, two loci would be involved; if ⅛, three loci. It was necessary, of course, to run controls along with the backcrosses, so 7 colonies from the hygienic Brown Line were tested, along with 7 colonies of the non-hygienic Van Scoy parental line. Also 8 backcrosses were made to the Van Scoy Line. . .
>
> The 29 backcrosses to the Brown Line broke up into 4 groups . . . 6 colonies showed complete hygienic behavior. . . None of the other 23 colonies showed hygienic behavior, but . . . 9 colonies left uncapped dead larvae in the combs . . . in contrast to the remaining 14 colonies, all of which retained dead brood in capped cells.

Thus it appears that two different recessive genes must be present and homozygous to produce hygienic behavior; one of them, *u*, is "for uncapping of cells containing dead larvae, the other a gene designated *r* for removal of dead larvae. A cross between the two opposite homozygous types would result in an F_1 which is non-hygienic. An F_1 queen would produce 4 kinds of backcross colonies in a 1:1:1:1 ratio. In this case a 6:9:6:8 ratio was obtained." This makes the behavior of offspring from a particular strain or cross predictable.

As to stinging behavior, "In the course of 98 visits to the Van Scoy colonies we were stung only once. In the course of 98 similar visits to the 7 Brown colonies we were stung 143 times." But it did not follow that all the hygienic backcross colonies stung, and the differences in this behavior among the backcrosses showed that it has a different genetic basis from that of hygienic or nonhygienic and that more than one or two loci are involved.

The results of most work on genetics of behavior in the higher vertebrates suggest that it is influenced by more genes, in a more complex way, than that of the invertebrate examples already cited and is less completely fixed in pattern. This is by no means to be taken as a rule, however. Dilger's (1964) study of *Agapornis*, mentioned on page 146, shows that certain acts or elements of behavior have a fairly precise genetic control and that the extent to which experience modifies them is limited, as if the genotype sets a boundary outside of which experience can have no influence.

Scott (1964) came to a different conclusion in his study of the genetic and other factors influencing the barking of dogs. The development of this behavior and its changes during individual life were investigated in cocker spaniels (notable for the amount of barking) and basenjis, which bark very little. Barking is "absent at birth, reaching a maximum later on, and declining thereafter." It is subject to the amount and kind of stimulation, and the threshold of response to stimulation is very different in the two breeds. So is the sound and the tendency to continue barking. As social behavior, barking depends on development of social relationships. The results of crossing the two breeds, and of backcrossing, show that inheritance is not a matter of simple dominance of either type but that there may be "a single major gene whose action is modified by complex interaction with either environmental factors or other genic systems. The data also suggest that a different gene system may be involved in controlling the amount of barking as compared to the threshold of response." Scott infers that the mode of inheritance promotes individual differences and the capacity of the individual to vary its behavior, both being adaptive advantages. This appears to be contrary to the idea that the genotype limits variability of behavior in an individual; actually it is not a contradiction but an extension of those limits. Even in man one could not say that the limits to variability in behavior have disappeared, only that in many aspects they are so remote that they are seldom noticed.

REFERENCES

Bastock, M. 1956. A gene mutation which changes a behavior pattern. *Evolution 10:* 421–439.

Caspari, E. 1958. Genetic basis of behavior, pp. 103–127. In A. Roe and G. G. Simpson (eds.), *Behavior and Evolution.* Yale Univ. Press, New Haven.

Coghill, G. E. 1929. *Anatomy and the Problem of Behavior.* Cambridge Univ. Press, New York.

Dilger, W. C. 1960. The comparative ethology of African parrot genus *Agapornis. Z. Tierpsychol. 17:* 249–685.

———. 1964. The interaction between genetic and experiential influences in the development of species-typical behavior. *Am. Zoologist 4:* 155–160.

Evans, H. E. 1962. The evolution of prey-carrying mechanisms in wasps. *Evolution 16:* 468–483.

Fuller, J. L., and W. R. Thompson. 1960. *Behavior Genetics,* Wiley, New York.

Harlow, H. H. 1958. The evolution of learning, pp. 269–290. In A. Roe and G. G. Simpson (eds.), *Behavior and Evolution.* Yale Univ. Press, New Haven.

Kessel, E. L. 1955. The mating activities of balloon flies. *Syst. Zoology 4*(3): 97–104.

Mayr, E. 1958. Behavior and systematics, pp. 341–362. In A. Roe and G. G. Simpson (eds.), *Behavior and Evolution.* Yale Univ. Press, New Haven.

Michener, C. D. 1964. Evolution of the nests of bees. *Am. Zoologist 4:* 227–239.

Pittendrigh, C. S. 1958. Adaptation, natural selection, and behavior, pp. 390–416. In A. Roe and G. G. Simpson (eds.), *Behavior and Evolution.* Yale Univ. Press, New Haven.

Rothenbuhler, W. C. 1964. Behavior genetics of nest cleaning in honey bees. IV. Responses of F_1 and backcross generations to disease killed brood. *Am. Zoologist 4:* 111–123.

Schmidt, R. S. 1964. *Apicotermes* nests. *Am. Zoologist 4:* 221–225.

Scott, J. P. 1943. Effects of simple genes on the behavior of *Drosophila. Am. Naturalist 77:* 184–190.

———. 1964. Genetics and the development of social behavior in dogs. *Am. Zoologist 4:* 161–168.

Sturtevant, A. H. 1915. Experiments on sex recognition and problems of sexual selection in Drosophila. *J. Animal Behav.* 5:351–366.

11 · GEOGRAPHICAL DISTRIBUTION

The processes and results of evolution differ geographically, as no population of organisms is universally distributed, nor are the factors that influence adaptation. Distance, by itself, can have biological effects. For instance, the proximity of males to females in organisms that reproduce sexually is in most cases essential, but distance can reduce or prevent breeding. Apart from this isolating effect of distance, the course of evolution in one area may be different from that in any other because the organisms are unlike and because the conditions to which they adapt also differ. For any population, whether of broad range or not, the manner of its evolution depends on local circumstances that it faces over a period of time. These are the simple physical conditions of the environment, as well as the biological relationships between it and other populations that impinge upon it as predators, prey, competitors, parasites, and so on. Beyond this, a widespread population is subject to factors that differ geographically in various parts of its range.

Much of the systematic study of plants and animals relates to physical and biological influences on local populations. The effects may be hereditary and mediated by natural selection or nonhereditary modifications occurring in ontogeny, perhaps differently in various ecological conditions. Evidently, then, if we pursued the relationship between geography and evolution to the limit, we would involve the greater part of biology; almost any biological phenomenon (even the color of hair or the manner of attaching a cocoon to a twig) may be subject to geographical difference. Here it will be convenient to limit the discussion to the broader results of evolution as shown by regional faunas, by the geography of groups of organisms, and the evidence of past regional changes and distribution of groups.

FAUNAL REALMS

It will serve our purpose here to give a sample of the characteristic faunas of different regions, adding that most of the principles illustrated by this apply equally well to plants. Recognition of faunal realms is based on the animals themselves, that is, on the distribution of particular species, genera, families, and larger groups as they are today or have been within historic times. Analysis of the causes of this distribution requires evidence on the past history of groups and of the lands they inhabit. As shown by both animals and plants of today, we recognize six geographical realms (Fig. 29):

1. *Palearctic,* in Europe and Asia north of the tropics, as well as the northwestern corner of Africa, including the Atlas Mountains.
2. *Nearctic,* North America exclusive of the tropics; this includes Alaska, Canada, United States, and the central plateau and mountains of Mexico.
3. *Neotropical,* Central America including the lowlands of Mexico, islands of the Caribbean, and all of South America.

Figure 29. Geographical regions of the world based on the distribution of land and freshwater faunas.

4. *Ethiopian,* Africa (with the exception of the Atlas Mountains mentioned), plus Madagascar, although this large island has a rather "incomplete" fauna and is sometimes treated separately.
5. *Oriental,* the tropical part of Asia south of the Himalaya Mountains and eastward through Sumatra, Java, Borneo, and the Philippines.
6. *Australian,* containing Australia, Tasmania, New Guinea, and all the islands of the Indonesian archipelago that lie east of Borneo, beginning with Celebes.

The fauna of any one of these realms contains certain distinctive groups of animals that are either limited to it (endemic) or nearly so; also it has representatives of other groups that may be found more abundantly in an adjoining region. It will contain some members of various worldwide or hemisphere-wide genera and families. A final category is that of highly distinctive groups (such as lungfishes), perhaps occurring as single species, whose nearest relatives are in a distant part of the world; rarely this applies to a larger group, as the marsupials of Australia and the Americas.

Where one faunal realm meets another, there is either a massive barrier across which few animals pass (as the Himalayan mountain system separates the Palearctic from the middle part of the Oriental, and as the Sahara and Arabian deserts separate it from the Ethiopian, or Bering Strait from the Nearctic), or there is a zone of mixing or overlap in which a portion of each fauna enters the other, but some groups meet their limits of distribution (as the blending of Palearctic and Oriental in southeastern China or the Nearctic and Neotropical in the eastern lowlands of Mexico).

The Palearctic and Nearctic realms together occupy the greater part of the land masses of the Northern Hemisphere, outside the tropics. The barrier of Bering Strait that separates them at the present time is neither very broad nor of long duration. These two realms therefore have much more in common than either of them has with any of the more southern faunal regions. Together, they are often called the Holarctic.

Palearctic. Among freshwater fishes (marine fishes are of no use in characterizing these realms) the most abundant Palearctic family is the Cyprinidae (minnows), closely related to those of the Orient; this is true also of the catfishes. Some perches and two species of suckers are Palearctic; their relatives are North American. Various genera of sturgeons and of salmon, trout, and their relatives, are shared by the Palearctic and Nearctic. One species of *Polyodon* (paddlefish) occurs in rivers of China, another in the central United States.

In Amphibia the salamander family Hynobiidae is almost exclusively Palearctic. Newts (Salamandridae) are represented equally well in North America. A single species of giant salamander (*Megalobatrachus*) has its nearest relative, *Cryptobranchus,* in the United States; likewise, the neotenic *Proteus* of European caves is perhaps most closely related to

Necturus of eastern United States. A couple of species of lungless salamanders of Europe are strangely isolated from a large number of genera and species belonging to that family in the Nearctic. The wide-ranging *Bufo* (toad), *Hyla* (tree frog), and *Rana* are present in both regions but much more numerous in the Nearctic. Of Palearctic reptiles the most distinctive is an *Alligator* in the Yangtse of China, related closely to that of the southern United States. Otherwise, the relatively few Palearctic reptiles are extensions of Oriental and Ethiopian families; the Old World vipers and wall lizards do not occur in the Western Hemisphere.

Of birds, most Palearctic species and genera are related to those of the Ethiopian and Oriental regions (as, for example, the weavers, starlings, and Old World warblers) or to North American groups (loons, grouse, waxwings, and creepers). Likewise, of mammals there are scarcely any distinctive Palearctic groups, except the mole rats and the pandas. All other families are shared either with the Nearctic or the Old World tropical regions.

Nearctic. Nearctic freshwater fishes are somewhat more distinctive of their region; for instance, the primitive holosteans, *Amia* and *Lepisosteus*, are characteristic of the southern United States. Freshwater sunfishes and black bass are exclusively Nearctic, as are the darters, a subfamily of perches. The Nearctic catfish family Ameiuridae barely enters the tropics, and the suckers are Nearctic except for two species that reach northeastern Asia. The numerous Nearctic minnows show a less obvious relationship with the Orient than do those of Asia and Europe.

North America and especially the eastern United States are rich in salamanders, primarily the families Plethodontidae, Ambystomidae, and Salamandridae. The more terrestrial among the plethodontids (without a larva) extend also far into the Neotropical, in upland forests where many kinds are arboreal. Frogs and toads of the Nearctic are generally of widely distributed groups, but one very primitive genus, *Ascaphus*, is found in wet forests of the Pacific coast and has no near relatives except *Leiopelma* of New Zealand; this is one of the most remarkable cases of animal distribution. The only exclusively Nearctic family of reptiles is the poisonous beaded lizards, or Gila monsters, with two species in northwestern Mexico and adjacent Arizona. The American alligator, as already noted, has a close relative in China. The turtles, lizards, and snakes are primarily of groups that range into the tropics or, like the colubrid snakes, are virtually worldwide.

The Nearctic bird fauna is very much like that of the Palearctic with the exception of a few groups that have their sources in the Neotropical, for instance hummingbirds. The turkey family occurs only in North and Central America. The families of mammals are mostly widely distributed, but the opossum is obviously related to those of the Neotropical forests. The American pronghorn (not a "true" antelope) and the so-called mountain beaver of the Pacific coast redwood belt constitute two exclusively Nearctic families. On the whole the family connections of most North American mammals are with those of the Palearctic, but a few are with the Neotropical.

Neotropical. The Neotropical fauna and flora are in many respects the richest and most interesting on earth. They contain a large number of endemic groups, a few special cases of discontinuity (that is, representatives of groups that are also known in some distant part of the world), and a number of animals that have their nearest relatives in the Nearctic region. The freshwater fish fauna is the largest in the world, containing primarily characins, several families of catfishes not known anywhere else, gymnotid "eels," perchlike cichlids, many cyprinodonts (killifishes), two osteoglossids, and a lungfish. About one tenth of the existing characins and cichlids occur in Africa. The lungfish, *Lepidosiren,* is most nearly related (although not closely) to *Protopterus* of Africa, and a small number of osteoglossids (an ancient family) are found in the tropics of the Old World.

Neotropical Amphibia, apart from families that are widely distributed, include the wormlike caecilians (also Africa and the Orient), lungless salamanders in Central America, the primitive pipid frogs (also Africa), and some exclusive small groups. Likewise, among reptiles a large turtle, *Dermatemys,* of Central America, the caimans, the boas (a subfamily of snakes), and two small families of lizards are Neotropical, but there are many other wide-ranging families.

The bird fauna is particularly rich. About half of the families are either exclusively Neotropical or may overlap into North America. Examples of these are the hummingbirds, flycatchers, motmots, manakins, cotingas, tinamous, and a group of five characteristically Neotropical families of insect-eating birds. All of these together number over 1200 species. We may add the unique hoatzin and rhea, or South American ostrich. Most other families of Neotropical birds are of wider distribution.

The Neotropical mammal fauna is more distinctive than any other except that of the Australian realm. It contains two families of opossums, the didelphids and caenolestids; only the Virginia opossum extends into North America. There are two families of platyrrhine monkeys, also anteaters, sloths, and armadillos (one of the latter reaches the southern United States), tapirs allied to those of southeast Asia, camels (llama, alpaca, and so on; the humped camels are in western Asia), about five exclusive families of bats, the caviomorph rodents (ten exclusive families plus the American porcupines, which extend into North America).

Ethiopian. The Ethiopian region, although of great interest, is somewhat less rich and varied than the Neotropical, and it has a smaller number of exclusive groups. The great size of Africa might lead us to expect otherwise, but two factors may account for this difference in the faunas of the Ethiopian and Neotropical. One is that Africa has not been isolated from Europe and Asia for a great length of time as South America was from North America; the other is that the area of dense tropical rain forest in Africa is considerably smaller than that of South and Central America, and, at the present time, a large part of Africa is either desert or semiarid, unable to support rich vegetation.

Ethiopian freshwater fishes include some exclusive families of catfishes and herringlike fishes, one lungfish, and two genera of primitive bony fishes, *Polypterus* and *Calamoichthys.* There are cichlids and characins, but not nearly as many as in the Neotropical, plus a few cyprinids, cyprinodonts, and some perchlike fishes. In Amphibia, caecilians are shared with the Orient and South America. The pipid frogs (as *Xenopus*) have relatives in South America; there is one exclusive family, but otherwise the frogs and toads are of wide-ranging groups. There are no salamanders.

Likewise, among reptiles are chameleons and gerrhosaurids (which have reached Madagascar), lacertid and agamid lizards, skinks, monitors, and amphisbaenids, as well as a couple of exclusive small families. Of snakes there are pythons, cobras, vipers, and the worldwide colubrids. Freshwater turtles are pelomedusids (also South America), and there are many crocodiles.

The bird families are mostly of wide distribution with a strong relationship to the Orient or, by extending northward or migrating annually, to the Palearctic. Hornbills, hoopoes, and bee eaters, for instance, occur across the Orient and beyond it. Several groups that are limited, or nearly so, to the Ethiopian are tree hoopoes, bush shrikes, widow birds, tick birds, buffalo weavers, and guinea fowl. There are also a few exclusive families: mouse birds, touracos, helmet shrikes, and the ostrich, hammerhead, and secretary bird.

Among mammals, the apes (gorilla, chimpanzee), monkeys (including baboons), antelopes and buffalo, elephant and rhinoceros all have relatives in the Orient, but there are many exclusively Ethiopian families: giraffes and okapi, hippopotamus, aardvark, hyraxes (these reach Asia Minor), golden moles, otter shrews, anomalurids, and four families of porcupine-like rodents. Zebras are conspicuous, but other members of their genus, *Equus,* are Palearctic. A species of hippopotamus reached Madagascar but became extinct there.

Oriental. Rivers and lakes of the Oriental region, including Sumatra, Java, and Borneo, have a rich and abundant fish fauna, the major elements of which are Cyprinidae (minnows) and their relatives, the loaches, and a great number of catfishes. Equally distinctive but less numerous are the labyrinthine fishes (ophicephalids and anabantids, of which one genus extends to Africa), channids and mastacembelids also reaching to Africa, and a few others.

Among Oriental Amphibia are caecilians, one or two salamanders that barely enter the northern edge of the region, a *Hyla* that is also marginal, but primarily ranids, rhacophorids, bufonids, pelobatids, and brevicipitids. Of reptiles there is an exclusive platysternine turtle; the gavials are also exclusive, as is the lizard *Lanthanotus* of Borneo. A few snakes are exclusive. There are many other lizards and snakes of widely distributed families.

Birds are numerous in many parts of the Orient but are primarily of groups shared with Africa, the Palearctic, and the Australian regions. The only exclusive family is fairy bluebirds; most pheasants are Oriental. In mammals, again, most families are shared with the Ethiopian region, a few with the Palearctic, and some bats with the Australian. Exclusive families include tarsiers, tree shrews, "flying lemurs," and spiny dormice. The Indian elephant and rhinoceros, the Malayan tapir, and apes (orangutan and gibbons) have geographical interest because of their separation from their nearest relatives elsewhere in the world. It appears that the Oriental tropics have for a long time served as a corridor for dispersal to and from areas to the west, north, and east by many animals, but only in certain cases can it be regarded as the place of origin of major groups: one such is probably the minnow family Cyprinidae and the allied Cobitidae and Homalopteridae.

Australian. In a sense the Australian region has been a last stopping place for several vertebrate groups that were unable to pass beyond, to the islands of the South Pacific. The fauna of this realm is sufficiently incomplete, compared with that of the Orient or Africa, to suggest that there was never a land connection between Australia and the rest of the world, but for a very long time a chain of islands extended, as today, toward Asia. These and the straits between them have served as a screen to pass some but block other groups of animals that might disperse eastward or westward. The freshwater fish fauna is extremely limited; there is a lungfish, *Neoceratodus*, and an osteoglossid, but otherwise fishes of the Australian region are peripheral (along shore) and enter fresh water from the sea, as, for example, various catfishes.

Among Amphibia certain families of frogs are numerous in favorable parts of the region but are all much more widely distributed elsewhere in the world. A river turtle, *Carettochelys* of New Guinea, represents an exclusive family, as do the pygopodid lizards. Chelyid turtles occur also in rivers of South America. Otherwise, geckos, agamids, skinks, varanids, elapids, colubrids, and pythons are all of widespread families. There are, incidentally, more poisonous than nonpoisonous species of snakes in Australia.

Australian mammals are frequently cited as a unique, primitive, endemic fauna which has survived in isolation from a time when marsupials were the most advanced mammals. This, however, is incorrect and misleading in several ways. The monotremes, of two families and two genera, are of course exclusively Australian and have no fossil record. Six families and 47 genera of marsupials are also exclusively Australian, other marsupials, of other families, being Neotropical. But there are 7 families of bats containing 21 genera, and of these no families and less than one third of the genera are endemic. One worldwide family of rodents, the Muridae, has 13 genera in the Australian region, most of them endemic.

Carnivora are represented only by the dingo, a species of *Canis* (dog), neither the family nor the genus being endemic; it is probable that the dingo was introduced by early man some thousands of years ago and was at that time partially domesticated. (Man's arrival there is now dated as not later than 26,000 years ago and at several times since, according to Tindale.) The rats and mice (Muridae), as discussed by Tate (1951) and Simpson (1961), reached New Guinea and Australia at various times, the earliest of which was Miocene or late Oligocene, and others Pliocene and Pleistocene. The older arrivals radiated adaptively in New Guinea and in Australia.

Concerning marsupials, the evidence now is that they and the placentals diverged in the Cretaceous from a common therian stock and that both were widespread by the late Cretaceous; probably it was pure chance that only the marsupials succeeded at about that time in crossing the island chain to Australia. Early marsupials did occur in Europe, North America, and almost certainly in Asia.

It should be emphasized, after this summary of the vertebrate faunas of the major geographical regions, that within each of the six faunal realms distribution is by no means uniform for any group of animals or plants. Species, genera, and often families have ecological limitations (or "preferences") confining them, perhaps, to only a small fraction of the whole area. Before the time when men began to change the ecology of the earth, probably the Oriental region was the most nearly uniform. Much of it was low, hot, and well watered but with seasonal variation in rainfall. There were ranges of hills and mountains reaching a high altitude in India, relatively cool elevations even in parts of Viet Nam and Borneo, and varied ecological conditions: bare rock at the summit and montane forest on Mt. Kinabalu; pine–oak–grass communities in the highlands of central Viet Nam; lowland marshes, heavy rain forest, savannas, and the semiarid to desert plains of western India.

In the Holarctic realm various ecological systems have been proposed for recognizing or describing the biological communities that characterize various parts of the area. These are distinguished primarily by their faunas and floras, which represent particular fractions of those of the whole region but which are characteristic of, or adapted to, a certain combination of physical conditions, especially length of annual growing season, upper and lower extremes of temperature, and amount and annual distribution of rainfall. It is unnecessary here to discuss these in detail, but a few generalizations can be made.

For instance, many groups of plants, insects, and birds found in the arctic tundra at sea level also occur farther south at higher elevations along mountain ranges or on isolated peaks where similar conditions prevail. On the equator in Africa, the Andes, or New Guinea, one meets alpine meadows above timberline at an elevation of 10,000 to 12,000 feet, and above these bare rock or snowfields. Sometimes, surprisingly, there are members of genera characteristic of the far north, for instance a small group of species of *Boloria* (butterflies) on Mt. Kilimanjaro, the Ruwenzori, and other highlands of Africa; other species of these butterflies occur in northern Eurasia and North America. Going south from the equator there is a corresponding decline in elevation of the timberline and alpine zones, until sea level is reached again at the latitude of Tierra del Fuego or the South Island of New Zealand.

At the other extreme, evergreen and ever-blooming rain forest is restricted to areas with a temperature continuously high enough for the growth of vegetation; it also requires an annual rainfall of not less than 70 inches, distributed fairly evenly throughout the year. In well-watered tropical countries, as eastern India, the eastern slopes of the Andes in northern South America, or the Caribbean slope of Central America, such a forest, with great richness and diversity of vegetation, may reach elevations of 7000 or 8000 feet. In tropical and temperate regions there often is a marked seasonal difference in rainfall: one wet and one dry season, or two short dry and two wet seasons, depending on change of the direction and location of wind belts.

Evidently, then, in any great faunal region there is freedom for much adaptive radiation of groups that occupy it, subject only to competition with others or to environmental limitations. In a general way, the adaptive diversity of a group that is endemic to a particular region (does not occur elsewhere) is proportional to the time during which it has lived there, and the relative number of such diverse but endemic groups may give a clue to the degree of isolation of a region. Australia and South America are the extremes in this respect among faunal realms, and the reason for this is the same in both: They have had very long histories of separation, as islands, from other continents, Australia ever since some time in the Mesozoic, if not earlier, and South America during nearly the whole of Tertiary time, from the Paleocene to late Pliocene.

A final note here concerns the duration of climates and climatic zones. Some 25,000 years ago, a mere moment in geologic time, the latest Pleistocene glacial period was at its height. The average temperature of the world was a few degrees (perhaps 5°C) cooler than now; the amount of water held frozen in continental glaciers was sufficient to lower the sea from 200 to 300 feet below its present level, thus connecting some islands with the mainland and extending certain continental margins. Many existing deserts were then well-watered lands, but the area occupied by rain forest and strictly tropical climates was undoubtedly less than today. Such conditions were exceptional, however. Throughout the greater part of past time there were no extensive glaciers, the sea stood a little higher than now, and tropical faunas and floras often reached much farther from the equator than they do today. It would mean little, then, to suggest that a given tropical region (say Oriental, or Ethiopian) as we know it today was the center of origin of man or of any particular group of animals or plants at a time some millions of years ago, for the existing limits of those regions certainly are not the limits that were present then; today tropical climates are more restricted than they were in the Miocene, for instance.

ISLAND LIFE

Biologically, at least, there is no real distinction to be made between islands and any other land, but important differences may be found in many islands, depending partly on their size and structure, partly on the distances by which they are separated from continents or from one another, and partly, of course, on the time during which they have existed as islands.

If a familiar peninsula, such as Florida, Baja California, or the Yucatan peninsula in Mexico, were to be cut off from the mainland by a water gap several miles wide, owing to submergence of the land between, no appreciable difference in fauna and flora could be expected for many thousands of years. For a few, and only a few, groups of animals there would be complete genetic isolation from their nearest relatives—nearly complete in the case of many others, little or none in still other groups that were capable of crossing the straits. Among vertebrates, the freshwater fishes might be expected to show evidence, after thousands of years, of subspecific and then specific differences, and so might certain mice and the more sedentary reptiles. More obvious perhaps would be the changes in relative numbers, with increasing rarity or extinction affecting one side of the gap (for certain species) and not the other. But these would be only minor changes.

Precisely such differences as these (slightly complicated by cold climate) are seen when comparing the flora and fauna of the British Isles with those of adjacent Europe, or of Ireland with Great Britain, or Nantucket and Martha's Vineyard with Massachusetts. All of these were connected with the mainland within the last 10,000 years and before that were buried under, or adjoined the edge of, a continental glacier. With recession of the ice they became habitable by cold-adapted mainland animals and plants, but at the same time, with the slow rise of sea level as a result of melting ice, they were cut off from the mainland through valleys of the former channel rivers that carried much of the runoff from the glacier. Obviously, then, the time of isolation has been brief, little genetic or adaptive change has been accomplished, but there are faunal and floral differences that consist primarily of reduction in numbers of mainland species in the islands. Snakes, for instance, had no opportunity to reach Ireland before it was separated, although some did reach Britain.

It is thus, generally, with coastal islands in which the environmental conditions remain essentially like those on nearby continents. Of course, such islands are on the continental shelf and have relatively shallow water between them and other land. On a far greater scale, as regards the richness of fauna and flora, are the large tropical islands of Indonesia, the Philippines, and Taiwan, some of which have had mainland connections at a time no more remote than the height of the last glacial period, say 15,000 to 20,000 years ago. (Sumatra, Java, and Borneo together had a land connection with Asia; Celebes and others farther east did not; Palawan probably was connected with Borneo, but the Philippines in general may have had only a brief dry-land connection with Borneo or have been separated by narrow straits; Taiwan was connected with southeastern China.)

Although there is evidence that Borneo was separated while Java and Sumatra still connected with Asia, it is interesting that "the Leopard now occurs (among these three islands) only on Java, not on Sumatra or Borneo; the Proboscis Monkey is confined to Borneo; the Banting (Wild Ox) is on Borneo and Java but not Sumatra; the Orangutan, a bear, and the Two-horned Rhinoceros are on Sumatra and Borneo but not Java; the Wild Dog, Tiger, and One-horned Rhinoceros are on Sumatra and Java but not Borneo; and the Elephant, Tapir and Serow are on Sumatra but not Borneo or Java . . ." (Darlington, 1957, p. 490). In view of the general distribution during the Pleistocene of all these and others on all the lands mentioned, there has evidently been a history of postglacial reduction in the faunas of what are now separated islands, but involving different species in each. Man may have had something to do with it, but so also may the sizes of the islands. The freshwater fishes are very numerous, closely related to or the same as on the mainland, but decreasing in numbers of species in some areas, such as eastern Java; a moderate number reach the Philippines.

Contrasting with these examples is Madagascar, lying some 260 miles off the east coast of Africa and of a size and structure comparable to Borneo, although the climate is at present drier. Between it and Africa is fairly shallow water, occupied by a small archipelago, the Comoros.

Distance alone is not enough to account for the very marked dissimilarity of its fauna to that of any other land; it appears never to have had a dry-land connection, at least not since the Mesozoic. (Connections with India or Africa have often been suggested, but the evidence is against this.)

The vertebrate animals of Madagascar, especially the mammals, are of fewer groups than those in adjacent Africa or in India, and many of them show evidence of long adaptive evolution since the time of their arrival. For instance, the lemurs, in three families and some ten genera, are not closely related to other lemurs but are presumably a natural group that reached the island early in the Tertiary as, probably, a single ancestral species. There is also an endemic family of insectivores, the tenrecs, of about ten genera, seven endemic genera of rodents (a subfamily), and seven endemic genera of viverrids (carnivores), the latter of which perhaps arrived on three or four different occasions. A small hippopotamus, not the same as now living in Africa, occurred there in the Pleistocene. But most of the major groups of African mammals did not reach Madagascar, as they would have done if a land connection existed. This indicates that the fauna originated from occasional strays that were compelled, on rare, accidental rafts of floating trees or logs, to drift across; the hippopotamus probably swam. Dinosaurs of Jurassic and Cretaceous age are reported from Madagascar; it is not certain that land was necessary for them, but a narrower strait may have been present at that time. Freshwater fishes, except a few that have some tolerance for seawater, do not occur there.

The kind of fauna represented in Madagascar is helpful in understanding that which is found on other islands much more distant from any continent, that could not have been connected by land at any time with the places from which their inhabitants came. New Zealand has two large islands with much ecological diversity, from temperate to cold climate, and is at least 1000 miles from the nearest important land, Australia. The islands themselves are of complex structure, like continents, and obviously have existed for a very long time, possibly always, without any close contacts. The fauna is small, but remarkable. The "living fossil," *Sphenodon*, or tuatara, is a lizard-like reptile, last survivor of its order, Rhynchocephalia, which was widespread in the early Mesozoic; this animal occurs today on a few small islands just offshore from the North Island of New Zealand. Another unique endemic is the Kiwi, sole member of its order of flightless birds. Present until recent times but destroyed by man were the moas, a family of very large ostrich-like birds, Dinornithidae. There is one genus of primitive frogs, *Leiopelma*, with three species; their nearest relative is *Ascaphus* of western North America. Of course, a moderate number of flying birds and some bats occur in New Zealand. The point demonstrated is that a thousand miles of ocean do not prevent the passage of a few animals that can neither fly nor swim, although they make it extremely unlikely, and millions of years may elapse between one such arrival and another. (Now, of course, the situation is wholly changed by man's introduction of many domestic and game animals that are altering the ecology of New Zealand.)

As far as geographical distribution of life is concerned, there is probably little reason to distinguish between islands of "oceanic" and "continental" types. The former term is used for those that seem always to have been separated from other lands by wide, deep water, and for islands that, like Kwajalein or Eniwetok, are built primarily as coral reefs or atolls, or, again, as the Hawaiian and Galápagos islands, grew up from the sea floor by volcanic extrusion. Many oceanic islands are combinations of these. New Guinea, on the other hand, like Madagascar, is of "continental" type and has been connected with Australia at times. Geologically it is like a continent, having mountain systems, river systems, long sequences of sedimentation and erosion, an old igneous base, and so on.

But the faunal and floral characteristics of an island depend upon its size (area), the time during which it has existed as a possible habitat, the ecological diversity of its surface, its relation to wind and sea currents, and its accessibility by ancient or recent land connections or by chains of islands along which some dispersal can take place, and so on, more than on its method of origin or geologic structure. Two additional examples are of special importance for their bearing on evolution.

When Charles Darwin, as naturalist on the voyage of the Beagle, visited the Galápagos Islands in 1835, he found that the giant land tortoises were of related, but recognizably different, species on most of the twenty islands. He noticed similar differences in mockingbirds, and later, although not while he was there, he realized that this was true of a group of finches he had collected (now placed in a subfamily, Geospizinae). Of the latter he wrote: "Seeing this gradation and diversity of structure in one small, intimately related group of birds, one might really fancy that from an original paucity of birds in this archipelago, one species had been taken and modified for different ends" (Darwin, 1845; see Fig. 3). The idea grew in his mind as he accumulated more evidence on variation in nature during later years, leading to publication of *The Origin of Species* in 1859. These islands are volcanic (the activity has not yet stopped), emerging from deep water on the equator, 600 miles west of the coast of South America, and their scanty fauna shows clearly that a few, rare visitors, arriving by chance over a period of perhaps 3 or 4 million years, initiated the various endemic groups now present. It is most unlikely, although some authors have suggested it, that the islands ever had any connection by land with South America or any other part of the world.

Simpson (1953) under the heading "Odds in the Hawaiian Sweepstakes" tabulated the probabilities (or improbabilities) of the arrival of ancestors of the various groups of animals and plants native to the Hawaiian Islands, using information from Zimmerman (1948). Presuming that these islands are about 5 million years old, the table is as follows:

GROUPS	ENDEMIC SPECIES	ESTIMATED ANCESTRAL SPECIES	COLONIZATIONS PER 100,000 YEARS	ODDS PER YEAR
Land mammals	0	0	0	Less than 1 to 5,000,000
Land birds	70	15	0.3	1 to 333,333
Insects	3722	250	5	1 to 20,000
Land snails	1061	25	0.5	1 to 200,000
Seed plants	1633	270	5.4	1 to 18,518

The distance to the nearest continent, North America, is 2000 miles; there are small Pacific islands south and west of the Hawaiian group, but these are generally separated by hundreds of miles of sea. Because of the size and ecological diversity of the Hawaiians and their ample rainfall in certain parts, much more adaptive radiation has been possible, apparently, than in the Galápagos, so a few arrivals have produced large numbers of descendant genera and species. Yet the estimated number of separate arrivals required to accomplish this is very small, and the intervals between them long, amounting at least to many thousands of years. There are no freshwater fishes, amphibians, reptiles, or mammals except such as have been introduced by man.

The honey creepers, family Drepanididae, are an endemic group of Hawaiian birds, all undoubtedly descended from a single species, which was perhaps an American tanager (Thraupidae); it was relatively short-billed and a honey and insect eater. From it "there have gradually differentiated both short- and long-billed seedeaters and fruit eaters. Plumage patterns and coloration have varied to a high degree" (Baldwin, 1953, p. 286). The line diverged into two subfamilies, with at least 9 genera and 22 species as now classified (see also Amadon, 1950); many more "forms" have been named. About one quarter of the species and subspecies have become extinct since the islands were discovered, as a result of human interference with ecology.

Island faunas illustrate several principles, but these may only be applied with caution because various groups of animals differ so much in their capacities for dispersal. Since continents have the greatest and most varied faunas because of the area, diversity, and stability, islands become populated from continents more frequently than the reverse. Sometimes the movement is fairly simple and direct, but more often it involves steps or chains of islands, and of course it may in many cases go in both directions. Dispersal of a particular group of land animals through a series of islands usually is subject to much delay, and chance plays a large part in their ability to cross the water gaps; therefore the islands nearest the place of origin receive the greater number of kinds and those farther away a gradually reduced number, until the limits of dispersal are reached by very few. This sort of dispersal has been called "sweepstakes," because the chance of success is small and the "winner" is unpredictable. The resulting pattern of distribution (that is, at a specific time, such as the present) may be called an "immigrant pattern" (Darlington, 1957, p. 484). But as an immigrant group undergoes local evolution and partial extinction in various parts of its total range, some of the species may disappear from islands here and there, and others become expanded or begin to spread in a reverse direction, so that the previous pattern is more difficult to recognize, and finally there may remain only scattered portions of it as the group declines and is succeeded by others; this disconnected and apparently meaningless distribution Darlington refers to as a "relict pattern." Relicts are older than first immigrants but may not be very old, as this depends on the nature of the group and the problems involved in its dispersal. There are, however, a few relict animals of great age, whose time and place of origin and dispersal are wholly unknown because all related species are extinct; such are the Australian lungfish, *Neoceratodus*, and the African and South American lungfishes, *Protopterus* and *Lepidosiren*, the New Zealand frog, *Leiopelma*, and reptile, *Sphenodon*, and of course the monotreme mammals of Australia and New Guinea. Others might be added to the list, such as *Ascaphus, Amia, Polypterus*, and so on, which are survivors of groups widespread long ago, but it is not always possible to say how widespread or how long ago. It is obvious, however, that islands sometimes furnish places of refuge for such relicts, whereas other equally authentic relicts live successfully in the midst of continental faunas.

CONTINENTAL DRIFT

Both the gradual uplift and sinking of land surfaces relative to the level of the sea have long been known and are well documented. In some areas connections between continents have been established and broken repeatedly by this means. Islands have risen out of the sea and have disappeared again. Parts of continents have sunk, making seaways, such as the broad Cretaceous sea over west-central North America, or the Sea of Tethys that separated northern Asia from India. To the geologist many of these episodes are as familiar as the Norman conquest or the Napoleonic wars to a historian.

Another idea has been developed in the past among geologists, sometimes supported, more often discredited, but currently receiving serious attention. This is a theory that some land masses may have had different locations in the remote past from those they occupy today. Looking at a map of the world, preferably a globe rather than a flat map, and imagining that continents are separate units, somehow movable like the parts of a Chinese puzzle, it is intriguing to imagine, for example, that South America might have fitted against Africa, making originally one continent, then slowly moved away to the west, opening a basin that is now filled by the South Atlantic Ocean. On looking closely at details of the two coastlines, one is startled to find that there is a remarkable similarity between the eastern bulge of South America and the great embayment of western Africa; this extends even to corresponding irregularities of the coastline of each. If the map shows depth contours of the Atlantic Ocean, we notice a mid-Atlantic ridge running northward almost exactly halfway between the two continents and continuing into the North Atlantic between Europe and North America. It parallels rather closely the coastal outlines of the continents, as if it had some meaning related to them.

It seems highly improbable to most geologists that a section of the earth's crust as huge as a continent could be moved, as a whole, for hundreds or thousands of miles, no matter how long it took to do so. Even if a mechanism for causing this were conceivable we would still need compelling, unmistakable evidence from a source other than a map that this must have happened.

In a few cases geographical evidence of present or past distribution of animals or plants has suggested that it would be convenient, in explaining unusual distributions, if separate continents had formerly been attached. Every such example must be subject to rigid examination to find out whether any other explanation is conceivable before we think of accepting a theory as drastic as that of drifting continents. Almost always biologists and paleontologists have been willing to admit that a group, for example common to Africa and South America, might formerly have had a northern hemisphere distribution, even if fossil evidence for this were little or none. In fact, there is much evidence that seems clearly contrary to the idea of drift in the animal life of these continents. The unique features of South America fauna have already been emphasized. There is no direct relationship between the South American and African monkeys or other mammals (porcupines are now thought to be independently developed in two major groups of rodents). Many (such as apes, giraffes, antelopes, sloths, and marsupials) found in one have never occurred in the other. The same is true with birds and almost all other animals, as long as we limit ourselves to groups that are known to have originated since, say, the Jurassic.

At this point a different kind of evidence begins to show. It is not yet extensive. South African Triassic reptiles are well known, especially the therapsids or mammal-like reptiles, which are represented there by many hundreds of species. In the last decade expeditions to western Argentina have disclosed a similar large fauna of Triassic reptiles in which apparently closely related species occur, although these have not been found elsewhere. The fauna as a whole is not the same, but South America and Africa had certain genera in common. It seems probable that North America was not the route by which they came.

More convincing still to the present writer is evidence on the primitive characin fishes of the fresh waters of South America and Africa, as given by Myers (1967). These fishes are not only the stem group of the large order Ostariophysi (minnows, carp, suckers, and catfishes) but they contain members that are anatomically more primitive in some ways than any other teleosts, and the origin of the group must go back to early Jurassic or Triassic. It is among these ancient and strictly freshwater characins, which do not occur even as fossils in the northern hemisphere, that we find close relationships between those of the two southern continents. Myers concludes that "characoids evolved in a Jurassic or even a Triassic southern continent, which soon thereafter split to form the beginnings of the South Atlantic Ocean."

A further discussion of the means by which land masses can drift over the surface of the underlying molten magma, at a rate of perhaps 2 or 3 centimeters a year, and the bearing of this upon the formation of the Great Rift Valley, the Red Sea, and the hypothetical Gondwanaland, is beyond the scope of this book, but there has been a recent awakening of interest, with many publications on the subject. See, for instance, an excellent symposium, Gondwanaland Revisited: New Evidence for Continental Drift, in Proc. Am. Philos. Soc., *112* (5), 1968.

DISTRIBUTION IN THE SEA

In some respects the distribution of organisms that live in the sea is controlled by ecological factors comparable to those that operate on land. The habitats occupied by various animals and plants may be determined by temperature, light, the nature of the substratum, or the presence of other organisms. But for the great bulk of marine life the water and its movements control dispersal and distribution. Even if animals are, at one stage of their lives, sedentary or attached to a surface, as the corals and many mollusks and echinoderms, nevertheless microscopic eggs are broadcast into the water and fertilized externally, and there are free-swimming larvae. Great numbers of animals and plants spend their lives drifting at the mercy of the current, or swimming, but without enough power to take a direction contrary to that of the moving water. These are the plankton. Many of them can rise or sink in the water, commonly approaching the surface at night and descending to where the light is weaker by day.

Evidently, then, the patterns of geographical distribution in the sea are determined by the relation of the organisms to the movement of masses of water, or to the shore, surface, or bottom of the sea. But something should be said here about the different habitats available for living. One is the bottom or floor of the sea, extending from the shore, where the tide alternately covers and exposes it, to the greatest depths, reaching more than 36,000 feet; the average is about 12,000. Animals living on the bottom, whether moving freely, or sessile, or burrowing in it, are *benthos.* The zone along the shore is *littoral,* and this is the most diversified of all habitats in the sea as well as the richest in benthos. It begins with an intertidal region, where organisms must in various ways adapt to a rhythmic change in the level of the water and usually to the powerful impact of breaking waves. They may occupy an open, exposed coastline, a sheltered lagoon or estuary, or a river mouth. The salinity, oxygen concentration, turbidity, temperature, and light in these circumstances vary greatly. The bottom may be of hard rock, or shifting pebbles, gravel, sand, or fine soft mud, and the fauna and flora differ accordingly.

Below the level of low tide the benthic habitat extends, on most coasts, outward for a few miles, descending gradually to about 100 fathoms, then dropping somewhat more abruptly into the deep abyss. The continental shelf thus marked varies in width, sometimes reaching 100 miles or more where the coast has been geologically stable for long periods (as off the southeastern United States), or so narrow that it scarcely exists at all (as along some parts of western North and South America). Usually light penetrates for a few hundred feet but is adequate for photosynthesis only in the upper part of this.

The floor beyond the continental shelf is not, as formerly supposed, a flat plain but has varied topography of valleys, ridges, submarine mountains, and deep depressions. It is beyond the reach of daylight, is little influenced by variations of temperature (remaining at or near 4° C), and the water moves slowly, not enough usually to stir the fine sediment of the bottom. Oxygen concentration is nearly always adequate for animal life. The pressure varies with depth and other factors, but the effective viscosity of seawater is fairly constant throughout; this latter factor is of major importance in connection with the swimming or drifting properties of organisms.

Above the bottom of most parts of the sea lie many thousands of feet of water. The deeper, dark, and cold abyssal zone is filled with many kinds of fishes and invertebrates, some of which are luminous, but there are no photosynthetic plants. The source of food is in part the plankton and the slowly falling residue of organic matter from the surface waters, but there is much direct predation. Over the abyss is an upper zone of surface currents, wave action, light, and varying temperature, controlled largely by the wind. Here live the pelagic fishes—the whales and porpoises and pelagic birds (albatrosses, petrels, auks, and others)—and the floating plants (Sargasso weed) and animals (Portuguese man-of-war), as well as the great mass of plankton, both plant and animal.

There is always a close relationship among temperature, current, and wind, and these have much to do with distribution of life in the sea. Water has its greatest density at 4° C. Therefore any water that has been cooled to that temperature sinks below water that is warmer (or colder). The surface water of the sea freezes at a few degrees below 0° C, depending on salinity and movement; at the freezing point water is less dense than at 4° C, and therefore ice forms at the surface. In the Arctic and Antarctic oceans there is a slow creeping of the cold (but not coldest) water outward along the bottom into the other ocean basins, thus adding continually to the great reservoir of abyssal water at 4° C. But meanwhile the surface waters of unfrozen seas are circulated by the action of constant winds: the "polar easterlies," the "prevailing westerlies" at middle latitudes, and the easterly "trade winds" in the subtropical zones of both hemispheres. The latter create and maintain an equatorial current toward the west at the surface across both the Atlantic and the Pacific. The Atlantic equatorial current is deflected northward along the coast of North America as the Gulf Stream; it crosses the North Atlantic toward Europe with the help of the westerlies, and then the water that is not diverted into the North Sea or Arctic turns southward again on the western coast of Europe and Africa, finally to complete the circuit. In the Pacific a similar diversion of the equatorial current toward Japan carries warm water up the coast of Asia, crossing toward Alaska, and then, considerably cooled, down the western coast of North America and Central America, to finish its circuit off the coast of Ecuador. A corresponding current in the southern hemisphere moves counterclockwise in the Pacific and the Atlantic.

Although these wind-created currents are of surface water, not more than a few hundred feet in depth, they nevertheless cause an enormous movement of warm water into cold parts of the world, and vice versa. Countercurrents, less well known, are found beneath or adjacent to these surface currents. In some places, as on the western coast of Africa or the coast of southern California or Peru, there is a strong upwelling of cold water from greater depths. Little by little, therefore, all water of the oceans is drawn into and moved in these currents; a continual mixing of dissolved oxygen and other substances, resulting in almost uniform distribution, takes place throughout the sea.

Obviously many of the smaller plankton organisms found in surface waters within reach of currents are carried by them around and around in a circuit through the oceans as generations succeed one another. Therefore it is characteristic of plankton, as well as pelagic fishes and other swimming organisms, to have an extremely wide distribution of species. It might be expected that the swimming larvae of various benthic or littoral animals along a coast would be dispersed by the moving water, and this is the case, but such dispersal may be limited in effect by changing temperatures of the water or by lack of suitable habitats for the adults when they become sedentary. On the western coast of Central America larvae of such organisms may in part be swept out to sea in the equatorial current that crosses the Pacific, but apparently littoral organisms cannot survive the journey of many months, as they transform before reaching a littoral environment; species and genera of the western Pacific are generally not the same as those of the eastern.

Although the adult eel, *Anguilla,* is a strong swimmer and comes without difficulty from fresh water to the sea, and thence to a point in the Atlantic Ocean southeast of Bermuda to spawn, its young larvae (leptocephalus) are feeble swimmers and are borne along slowly by the water, first toward the west and then northward in the Gulf Stream. After a year, the larvae of the American species, *A. rostrata,* make their way to river mouths, transform to elvers, and swim upstream, but the European eels, *A. vulgaris,* require three years before they can enter the rivers of Europe.

A few cases are known in which marine animals of one species, or of closely related species, occur in north-temperate or subarctic latitudes and also in south-temperate or subantarctic waters but do not occur in tropical zones between. This is called bipolar distribution, although it does not imply that organisms reach either pole. A shark (*Lamna cornubica*), a tunicate, a pteropod mollusk, and a gephyrean worm exemplify this. Several kinds of copepods and other plankton live in surface waters of the northern and southern hemispheres, but in deeper water within the tropics, thus not being in fact bipolar but appearing so if only the surface water is investigated. An earlier connection of populations through deep, cold tropical waters may account for some of the present cases of bipolarity (with extinction in the zone between the northern and southern members); it is also possible that a former worldwide distribution at a time when the average temperature of the water was somewhat lower than it is now may have been followed by extinction in the tropical surface waters, leaving some populations separated.

Barriers to the dispersal of marine organisms are of two categories, one being the virtually absolute barrier of the shore, where the ocean stops, and of the atmosphere, into which no marine organism normally penetrates for more than a few seconds; the other is of barriers that are incomplete or perhaps of a gradual nature.

One of the latter is that of temperature, which may exclude warm-water species from areas that are cold or vice versa. Obviously some penetration in both directions is possible; usually the critical factor is the effect of temperature on breeding activity or on the survival of eggs or larvae. Another incomplete barrier is provided by reduction or increase of salinity. In the mouths of rivers many marine organisms enter part way. Fishes of several families (for example, the striped bass, *Roccus lineatus*) come from the sea into fresh water for spawning, and there are, especially in tropical waters, marine sharks, rays, and other kinds of fishes that habitually travel up big rivers for many miles from the sea. Organisms that cross such barriers of salinity with little difficulty are called euryhaline.

Another significant limiting factor in the sea is light. Of course the photosynthetic plants, including phytoplankton, cannot live successfully at a depth greater than the light required for photosynthesis will penetrate.

Differences of pressure at various depths in the water probably are overcome gradually and do not constitute absolute barriers, but presumably the distribution of animals at great depth is controlled by factors other than pressure.

PAST DISTRIBUTION OF PARTICULAR GROUPS

The present distribution of a group of animals represents its geography at a given moment of its history. In only a few of the better known groups is it also possible to show approximately its distribution from the time when it first appeared through the millions of years until today. This naturally requires that we know, from fossils of determined geologic age, the places where the animals occurred and, from deposits of sedimentary rock in other areas, where they did not occur. This information is more readily available in reference to horses, for instance, than it is for frogs or any special group of birds, or indeed any other animals whatever. Enough material must be obtained not only to identify and describe in some detail the species concerned but to make the age clear, usually by means of associated fossils of many other kinds as well as the stratigraphic position. (Sometimes we also have a close age determination in years from any of several methods, but this is not essential for judging the sequence in which fossils occur.)

The earliest known record of the horse family is from the early Eocene in both Europe and North America, where the little *Hyracotherium* (= *Eohippus*) was represented by a number of species (Fig. 30). This indicates quite certainly that a land connection was present at that time between the Old World land mass and North America, by way of the present Bering Strait. Later in the Eocene, the European stock gave rise to several primitive horselike genera, all of which died out as the epoch came to an end. In America, on the other hand, a different sequence of genera came in the mid and late Eocene, continued in the Oligocene, and

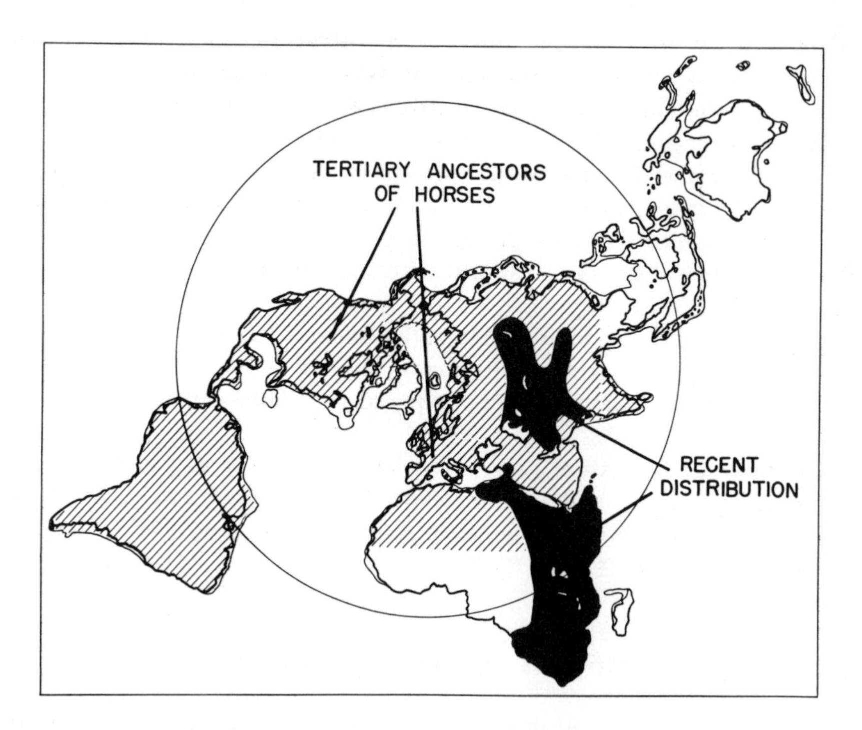

Figure 30. Tertiary and Recent distribution of the horse family. (Redrawn after Matthew.)

by the Miocene had produced various divergent lines of larger animals, most of which had no bearing on later horse evolution. Some, belonging to *Anchitherium* and, later, to *Hypohippus,* crossed again to Asia in the Miocene and Pliocene; these were still rather primitive, three-toed horses. Meanwhile another line of North American horses began to develop modern characters of the molar teeth, increase in size, and further reduce the lateral toes (*Parahippus-Merychippus*), producing *Hipparion* of the Pliocene, a genus that again spread from America into Asia, Europe, and Africa. Its African descendant, *Stylohipparion,* finally died out in the Pleistocene, the last of the three-toed horses.

Pliohippus, coming in the early Pliocene of North America, was the first genus of one-toed horses. It spread into the Old World, and in the latest Pliocene to South America, which previously had been isolated by a water barrier. The modern *Equus,* containing all living horses, zebras, and wild asses, arose from North American *Pliohippus* in the Pliocene, and also entered South America, after *Pliohippus* did. The South American descendants of the latter, as well as members of *Equus* in that continent, became extinct near the end of the Pleistocene. Meanwhile there was a migration, or more than one, of *Equus* into Asia, Europe, and Africa, where it remained abundant to the present time. Since the beginning of human history its distribution has become discontinuous, and it has been domesticated from at least two wild stocks. In the western hemisphere following the close of the Pleistocene all native horses, for unknown reasons, became extinct.

This brief version of the geographical history of horses illustrates a number of points that are significant in evolution of land animals.

1. When animals that are very closely related to each other, as shown by their anatomy, are found to have been present at the same time in widely separated parts of the world, we may infer with confidence that there was no barrier to their movement in one or both directions between the two localities.

2. The most effective kind of barrier to movement of terrestrial animals (especially mammals) is a sea between continents. Such a sea existed between South and North America during all the time from the appearance of the earliest horses until the late Pliocene; likewise a sea barrier has been present from time to time, as it is present today, between Asia and North America, the Bering Strait.

3. The center of the most extended, continuous evolution of horses was North America, although no native horses survived in the western hemisphere to the time of arrival of Europeans, who brought Old World domesticated horses with them.

4. The discontinuous distribution of Recent wild horses and zebras is accounted for by extinction in the areas between those they now occupy, just as the absence of native horses from the western hemisphere and western Europe is accounted for by their extinction in the Pleistocene, or later.

The story of the camel family, Camelidae, is like this but less completely known; it illustrates essentially the same principles (Fig. 31). So do those of the tapirs, rhinoceroses, and proboscideans (elephants and mastodonts).

In contrast to the fairly detailed picture of geographical history in some groups that have a good fossil record is the vague, incomplete history inferred from present distribution of many groups that have little or no fossil record. It is simple enough, of course, to say that hummingbirds originated in the Neotropical region, because all the kinds that are known live there, only a few of them migrating to the Nearctic for the summer breeding season. Although there is virtually no fossil record of platyrrhine monkeys, the same can be said of them, in full confidence. But to give any details is at present impossible. The following examples will show something of the way in which a little light may be shed on the subject from the present distribution of groups that are not limited to a particular, well-defined region.

The vast majority of Salientia (frogs and toads), say 190 genera and 1330 species (figures mainly from Darlington, 1957), live in the tropical parts of the world. The Neotropical realm exceeds in species (750) but

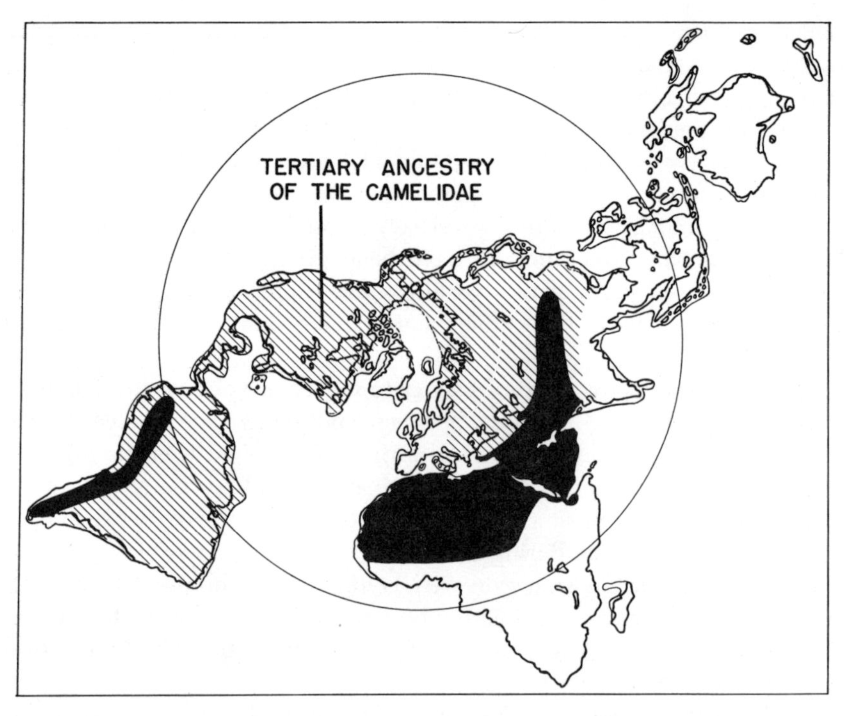

Figure 31. Tertiary and Recent distribution of the camel family. (Redrawn after Matthew.)

approximately equals in genera (94) the Ethiopian and Oriental combined (580, 96). The Australian region, from Celebes to Tasmania and the Solomon Islands, is not impoverished, but its frog fauna is restricted to 5 of about 14 families. The north temperate regions of both hemispheres receive relatively few species of 8 families, the south temperate even less. An exception to all this is seen, however, in the 3 highly successful genera *Bufo* (common toads), *Hyla* (tree frogs), and *Rana* (the more familiar frogs), which are almost cosmopolitan and number about 900 species together.

A recent classification of toads, family Bufonidae, gives 7 genera with about 145 species. *Bufo*, 125 species, is nearly worldwide but does not reach Australia, Polynesia, or Madagascar. The other genera are Oriental (4) and African (2). We might infer from this, perhaps correctly, that the family originated in the Orient, which is the area of its greatest present diversity, that it spread at an early time to Africa, but that only the versatile genus *Bufo* managed to escape beyond the limits of the Old World tropics. But 4 additional genera, and *Bufo* itself, are known as fossils in Europe (Oligocene to Pleistocene), and another (Miocene) is reported from South America. Assuming that all these fossils actually are

Bufonidae, two other hypotheses concerning the geographic history of the family could be constructed that are as plausible as the first. If we now add that studies of structure in toads are thus far inconclusive about the course of evolution in the family, we are left with this: *Probably* the center of its origin was in the main land masses of the Old World.

The large family Ranidae offers a clearer case than Bufonidae, because distributional and systematic data are less ambiguous. Three genera, including *Rana,* are found as fossils in the Miocene of Europe. Four of the six subfamilies are primarily or wholly Ethiopian. One subfamily (Cornuferinae), of 10 genera, spreads from the Orient to the Solomon and Fiji Islands. The subfamily Raninae contains 7 African or Oriental genera as well as the immense, nearly worldwide genus *Rana* (400 species). The latter failed to penetrate South America except for one species, *R. palmipes,* which probably arrived there no earlier than the late Pliocene, when the Central American corridor became connected. Thus an Old World tropical origin of the Ranidae is as clear as any fact in animal geography. Incidentally, the sharp restriction of one subfamily, Mantellinae (5 genera, 51 species), to Madagascar emphasizes the relatively complete isolation of this island with respect to animals that cannot tolerate seawater.

REFERENCES

Amadon, D. 1950. The Hawaiian honeycreepers (Aves, Drepaniidae). *Bull. Am. Mus. Nat. Hist. 95:* 151–262.

Baldwin, P. H. 1953. Annual cycle, environment and evolution in the Hawaiian honeycreepers (Aves: Drepaniidae). *Univ. Calif. Publ. Zool. 52:* 285–398.

Darlington, P. J. 1957. *Zoogeography. The Geographical Distribution of Animals.* Wiley, New York.

Darwin, C. 1845. *Journal of Researches.* 2nd ed. J. Murray, London.

Myers, G. S. 1967. Zoogeographical evidence of the age of the South Atlantic Ocean. *Studies in Trop. Oceanog., Miami 5:* 614–621.

Simpson, G. G. 1953. Evolution and geography. Published as a Condon lecture by Oregon State System of Higher Education. (Included in G. G. Simpson collected essays, *The Geography of Evolution.*)

———. 1961. Historical zoogeography of Australian mammals. *Evolution 15*(4): 431–446. (Included in G. G. Simpson collected essays, *The Geography of Evolution.*)

Tate, G. H. H. 1951. The rodents of Australia and New Guinea. *Bull. Am. Mus. Nat. Hist. 97:* 183–430.

Zimmerman, E. C. 1948. *Insects of Hawaii,* Vol. 1, Introduction. Univ. Hawaii Press, Honolulu.

12 · TIME AND THE GEOLOGICAL RECORD

NATURE OF THE RECORD

Our knowledge of the multitudes of species living today, in all their activities and interrelationships, amounts only to a glimpse of a passing scene. Extensive study of the existing kinds of animals and plants can show us what survives of the products of evolution but not, directly, the stages or the course of it. Were we in possession of a full record of life of the past, biological science would be unthinkably more complex than it is.

As an archeologist reconstructs, from imperfect relics occasionally preserved, something of the nature and mores of ancient people, so the paleontologist obtains from fossils, and the sediments containing them, dim views of far earlier scenes in the life of animals and plants. These he fits together into a steadily growing history, but the record as a whole will always lie beyond our reach.

In riverside cliffs, in the gorge of the Niagara or the Colorado, and in roadside exposures in many parts of the world one can examine rocks formed of layers of what was at first waterborne sediment. Finding that limestone, shale, or sandstone frequently contain ripplemarks, shells, silicified wood, leaf impressions, and now and then teeth and bones, we become aware that processes now in action in streams, lakes, and seas have been working for ages past. Seeing numerous layers of rock lying in succession upon one another, we can infer that each new deposit was placed upon the one beneath and in turn supports the next to be formed over it. The lowest member of a series should then be the oldest and the uppermost the youngest; only rarely has there been a later disturbance so complete as to create a problem in recognizing the actual sequence.

On the other hand, the ancient sedimentary layers are rarely left wholly undisturbed. Their position might at first have been horizontal, but commonly they have been tilted during subsequent time—either slightly, as in much of Kansas, or at steeper angles, as often seen on sea cliffs along the California coast. They may be gracefully folded, as if they were layers of carpets wrinkled by pushing the edges together (cliffs of the Allegheny and Susquehanna rivers). They may show cracks (faults) at an angle to the bedding plane and displacement up or down along these faults; this is especially common in the mountain states of the western United States.

Often it is possible to see where erosion once cut away part of an older series of sedimentary deposits but new layers were later added upon the eroded surface, perhaps in a plane that differs from the original one. This nonconformity in the sediments indicates a discontinuity in time, when deposition stopped and erosion removed an undeterminable amount; the time lapse may be from a few to millions, or even hundreds of millions, of years.

Sediments when originally laid down are in some cases nearly or completely hardened, as in the accumulation of some limestones, but usually the sediments are clastic (of particles), later consolidated to rock under increasing weight or by precipitation of a solid matrix among the particles. Sedimentary rock itself may at any time, if it undergoes tremendous pressure in the course of mountain building, or a degree of heating sufficient to cause softening and rehardening, become changed in texture. Sandstone is transformed to a crystalline quartzite, shale to a much hardened slate, limestone to a compact and somewhat crystalline marble. These are metamorphic rocks. The fossils in them are sometimes preserved but may be destroyed.

To these two categories of *sedimentary* and *metamorphic* rock a third, *igneous,* should be added. This is the kind that has been molten and became solid by cooling. If the cooling is rapid, as in the lava emitted by a volcanic eruption, the resulting rock may be bubbly or glasslike but has no crystal structure. But in igneous rock that solidifies far beneath the surface and therefore loses its heat very slowly, crystals form, as in granite.

In the course of an hour's walk it may be possible to see, depending on the manner of their surface exposure, rocks that were formed in one particular age, or perhaps in several, but there is no exposure showing a full succession of geological formations of all ages. Perhaps in the bottom of the most persistent ocean basins there may be an approximately complete record of what happened there, in the form of continuous deposits, but it is not likely to become available to us or to reflect much that happened elsewhere.

Even in the cores of oil wells drilled through thousands of feet of sedimentary rock the succession is not complete, because of the absence of any deposition during certain times in the past at a given locality and the removal of some deposits by erosion at such times. The tremendous wall of the Grand Canyon in Arizona shows more than 5000 feet of ancient rocks ranging from the Vishnu schist, one of the oldest metamorphic formations, up to Permian limestone at the rim, but this is by no means a continuous record, because rocks of several intervening periods are absent in that area.

Thus it is only by correlating the known formations of one region with those found in another, and filling the gaps in one series by the locally more complete records of others, that we have gradually established for the world as a whole a consistent "timetable" (Fig. 32). It is essential to this correlation that the characteristic fossils of each formation (usually this means a group of strata) in each area be known, for in a given formation the fossils are in some respects different from those in any other, above or below. Certain fossils are so distinctive of a given formation and are so likely to be found wherever that formation is exposed that they are used as "index" or "guide" fossils. This is especially true of widespread, common, easily recognized marine invertebrates; for example, various species of the "grain-of-wheat" shells, genus *Triticites* (Foraminifera), characterize and are limited to rocks of late Pennsylvanian and early Permian age, whereas blastoids (echinoderms) of the genus *Pentremites* are good indicators of Mississippian and early Pennsylvanian formations. Some, but not many, vertebrates can be used in a more limited way; usually they are too scarce and unevenly distributed to be dependable, although when obtained they are safe indicators; for instance, phytosaurs are known from late Triassic rocks in various parts of the world.

This use of fossils for the recognition and correlation of sedimentary rocks has nothing directly to do with evolution, except that as time passed the organisms that lived in a particular situation disappeared, changed, or were replaced; it makes no practical difference whether those that precede and follow are in any way related or not. They only need be recognizable. Thus, regardless of evolution, it was long ago possible to work out the exact sequence of rocks in terms of relative ages and index fossils; this has been a difficult task because of its magnitude, and it still continues as a subject for numerous studies, but the means for doing it have long been available.

The original recognition and naming of each geological period was the result of observations on rocks exposed in one or more localities where the beginning and end of a particular system of rocks may have been marked by a widespread or large-scale nonconformity. Frequently this interruption of the record was associated with evidence of mountain building, uplift of land, metamorphism, and other disturbances of the formerly more even deposition of waterborne sediments. But the geological events recorded there may not have worldwide implications. The ending of a long period of sedimentation by an episode of uplift and erosion may or may not be apparent in other regions than those in which it was first described. Nevertheless, certain extensive changes occurred in what seems like a rhythmic sequence to an extent great enough to bring land surfaces below sea level or elevate former lowlands to mountain ranges.

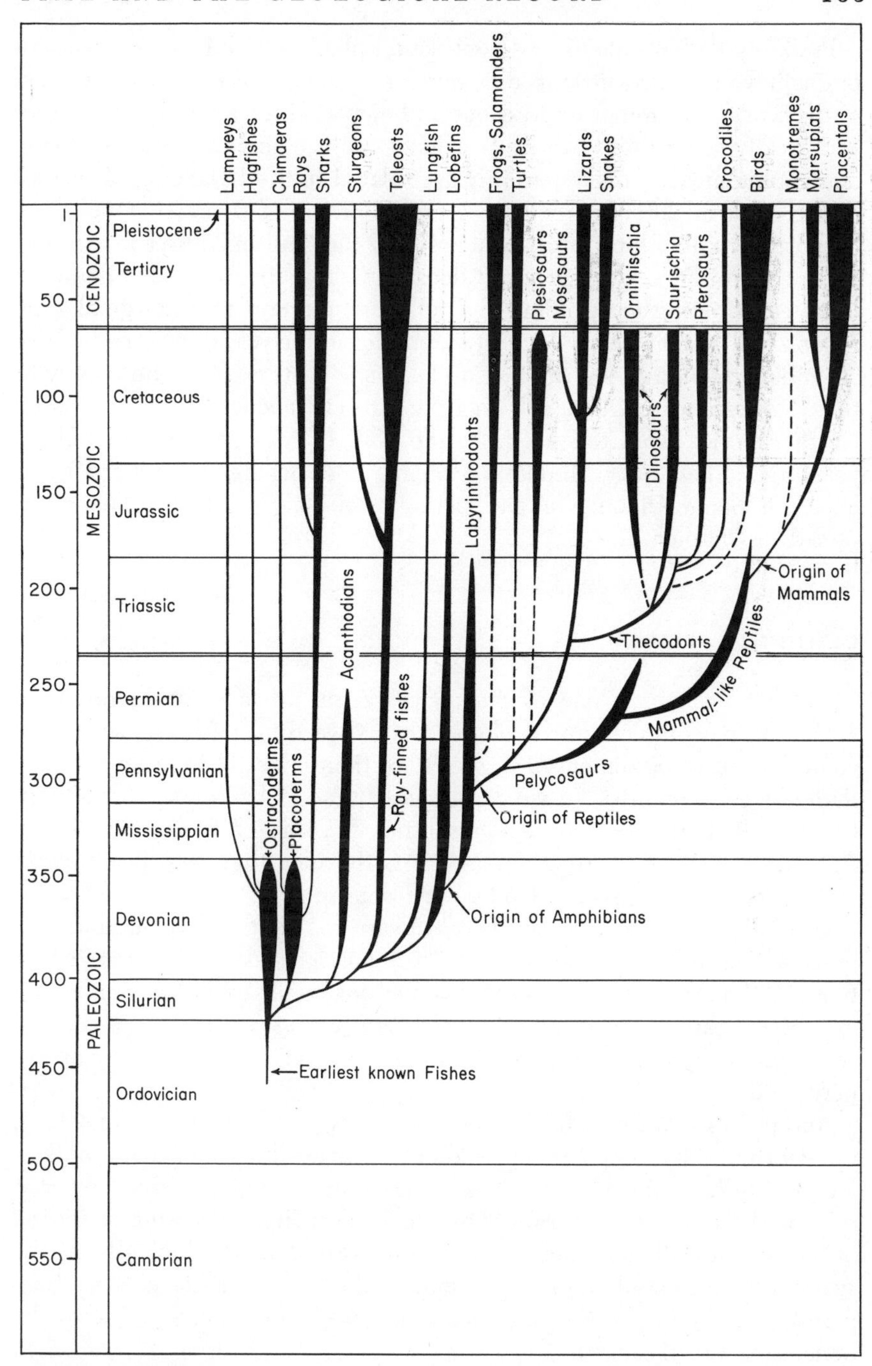

Figure 32. Evolutionary relationships of the major groups of vertebrates. Figures on left give age in millions of years. Not indicated on the chart is pre-Cambrian time, which was several times as long as that which is shown.

Probably the continents and continental platforms and the great ocean basins have changed little in the course of geologic history in most parts of the world, but minor depressions and elevations of a few hundred feet and locally some thousands of feet have taken place many times. These diastrophic movements apparently are related to the balancing of weight of the crust in different regions. A region of broad depression (geosyncline), gradually filling with sediment (say the Gulf of Mexico), accumulates weight and sinks lower as it does so, finally after tens of millions of years contributing to a folding and upthrust, perhaps in a marginal zone around it. The Himalayas, highest mountains in the world, now rise where 50 million years ago lay the Sea of Tethys, a waterway continuous with the Mediterranean. Older still was the late Cretaceous Sea that crossed North America from the Arctic to the Gulf of Mexico some 80 million years ago. The Rocky Mountain system stands where its western shore once ran. Some evidence of the possibility of continental drift was mentioned in Chapter 11.

GEOLOGIC TIME

Time is one of the major factors in any consideration of evolution, paleontology, or geology. Just as we think of space as having continuity in three dimensions, which are themselves abstractions, rather than objective realities, so we can think of time as having a kind of continuity called "duration" and representing in a sense a fourth dimension. But this, too, is an abstraction, and so is the idea that time flows like a stream regardless of events. Under conditions in which we have no knowledge of the succession of events, no awareness that one happening has preceded or followed another, we lose consciousness of the passing of time. Without the information that something occurred before something else, there can be no recognition of time, just as we cannot recognize space unless two or more points or objects having different locations are perceived.

And just as space may be measured by units, whether they be footsteps, the width of a hand, or the repetition of other arbitrarily selected units of length (miles or meters), so time is measured by units of duration that may be rhythmically recurring natural events, such as the rising or setting of the sun, or artificially selected units that also occur rhythmically, as the swinging of a pendulum or movements of the hand of a clock. Thus time partakes of the qualities of a dimension in ways quite comparable to the dimensions of space.

It follows from this that both space and time have meaning only insofar as we perceive them or can devise means of recording the extent or duration of them. Beyond the limits of our ability to perceive objects, that is, at a distance of a billion light-years or more, it becomes meaningless to say there is or is not space, for we can neither perceive nor measure it. We may imagine it, but this is only a projection of previous experience. On the scale of time, we can have no real basis for knowledge of a time prior to the earliest events of which the earth or the universe carries an intelligible record. Such an expression as "the beginning of time" is meaningless, although in the other direction it is possible to imagine a time scale going into the future for as far as we can pile up units of duration or events that succeed one another. These, of course, become increasingly uncertain (improbable) the farther they are detached from our experience, until they too fade gradually into the unimaginable.

As soon as it was established that the lower strata in a series of sedimentary deposits were older than those above them (Hutton in 1795, Lyell in 1830, and so on), there was a clear basis for putting in order a table of succeeding events in the history of the earth, as far as such deposits were accessible and as far as correlation of sedimentary rock in different localities could be made. The actual time required for this history, or for any significant part of it, remained unknown until recently, although more or less plausible estimates were made by many geologists. Therefore the geological timetable has been, and still in part is, relative rather than "absolute" in expressing duration. In a practical sense, this makes very little difference to the paleontologist. For most purposes it is sufficient to express time in eras (Mesozoic), periods (Cretaceous), epochs (Eocene), or smaller divisions of these (Albian, middle Eocene, and so on). Thus the major units of the geologic table, most easily remembered from earlier to later (because this was their actual occurrence), must be learned by any beginning student of evolution or paleontology (see Fig. 32).

The geologic table not only expresses a time sequence, uninterrupted for as long as geological information permits us to infer any past events, but it also (and separately) expresses a sequence of rock units, that is, the actual strata corresponding to the periods or epochs that are recognized. Rock units and times are often confused with each other in referring to them by name. Rocks are properly classified in systems, series (or stages), formations, and zones. The name of a system is the same as that of a period (for example, Permian system), but the divisions of systems are given more local, often geographical names (Missouri series, Niobrara formation—preferably with a descriptive term for the rock itself: Niobrara chalk), and if need be, still more subcategories.

Thus the broad outlines and many details of the succession of events and of the forms of life as shown by fossils were well known long before we had any precise knowledge of the absolute time scale in years. It was obvious by then that animals of the Pleistocene, some of them strange, extinct creatures (giant ground sloth, saber-toothed cat, and mastodon), were prehistoric, although certain species and genera of that time have persisted to the present. Far more remote, and utterly beyond the experience even of the earliest possible "caveman," were dinosaurs, the flying pterosaurs, the whalelike ichthyosaurs, the ammonites, and other characteristic animals of the Mesozoic. But a much greater leap than this was needed to take one back to the Cambrian, when all known animals were invertebrates, almost all of these marine, and there were as yet no insects, no fishes or higher vertebrates, and no trees, flowers, or seeds in the world.

Stretching into ages many times more ancient than the beginning of the fossil record lay the eras called Algonkian and Archean (collectively pre-Cambrian time). Here and there rocks of these ages are exposed, either igneous or highly metamorphic as a rule, but even where they have been least modified geologically they show little or no direct indication of life. The diversity of invertebrate phyla in the Cambrian (Fig. 33) strongly suggests that an immense span of time had been passed in the earlier evolution of small, soft-bodied organisms that left no actual fossils. Probably the origin of life (which we may think of as a prolonged process rather than an event) took place at a period three or four times as remote as the Cambrian.

Beyond all this was the beginning of the earth, the arrival of temperatures and circumstances in which there could be running water, rain, erosion, the deposit of sediment in layers, and the gradual initiation of geological activity—the slow emergence of a record that we can read today.

It was possible, in ignorance of the exact rate of sedimentation and other processes, to estimate that the entire geologic history of the earth took 100 million years or even less (as was believed at the end of the nineteenth century), but unless there were evidence resting upon a process that has a known and fixed rate, we have no reason to choose between 100 million and, say, 10 billion years for the same history. Estimates that one might suppose wholly reliable, such as the age of the oceans based on the rate of accumulation of salt, fall far short of accuracy because it is now known that this rate has varied greatly, being relatively rapid in times like the present when uplift of land increases both mechan-

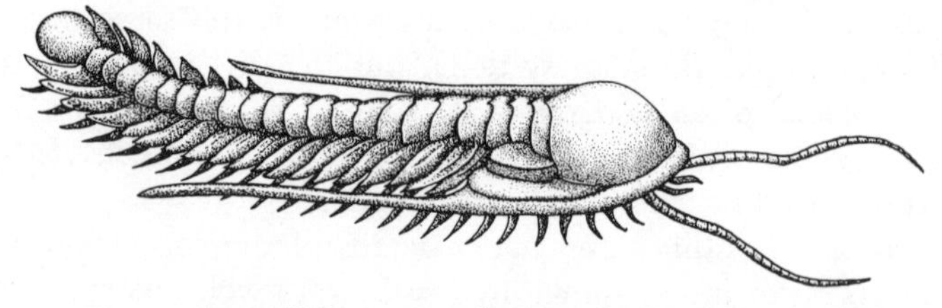

Figure 33. Trilobite, *Paradoxides harlani,* about 18 inches long. This was one of the largest animals of Cambrian time.

ical and chemical erosion, but slow in many other periods when land was low and bore few mountains or highlands.

There are now several methods available for measuring the absolute age of samples of rock in years, independently of their localities or positions in the rock series. These methods are based on the discovery that a radioactive element disintegrates at a determinable and constant rate, producing one or more other elements as it does so. The proportion existing in a sample of rock between the radioactive source still present and the product not yet completely formed can therefore give a close approximation to the age, from the time the sample first solidified with its existing mineral content. One of the best-known examples is the production of lead by uranium. This is not a chemical reaction; it is disintegration, at a fixed rate, of uranium atoms in a purely random manner. That is, the breakdown of any given atom may happen 10 minutes from now, or in a couple of centuries, or not for half a billion years or more, but it will happen.

Because the process is random and unpredictable as far as particular atoms are concerned, the rate is most easily expressed by stating the time in which one half of any given sample will be disintegrated, the "half-life" of the element. There are two isotopes of uranium (forms with different atomic weights and half-lives), each giving birth to a particular isotope of lead:

U^{238} (half-life 4498 million years) produces lead (Pb^{206}) + helium + heat.
U^{235} (half-life 713 million years) produces lead (Pb^{207}) + helium + heat.

Helium, being a light gas, escapes and cannot be measured, but the lead is retained in the rock. There are several difficulties in establishing exact ages of rocks by this method, but they are being overcome and the results are far more dependable than were the simpler estimates made earlier. One difficulty is in precision of measurements in the sample, for the two isotopes of lead that have been derived from the two isotopes of uranium must be measured, but the ordinary lead must be disregarded because it was not of radioactive origin. The amounts of all these may be extremely small. Minerals containing uranium cannot be found everywhere, but are largely limited to regions of igneous rock. Even if measurable amounts were available in sedimentary rocks (as in igneous pebbles included in the sediment), they could not give a clue to the time of deposition, for inasmuch as the pebbles originated elsewhere, the time of their first formation has nothing to do with their removal, transportation, and redeposition in later sediments. Therefore, determination of the age of sedimentary rocks, using the uranium–lead method, must be an indirect estimate based on the relative positions of the sedimentary rocks and of underlying or overlying igneous rocks in which the measurements are made. If the igneous lie upon, or have intruded into, sedimentary strata, then they must be more recent than the latter; in turn, they are older than sedimentaries that cover them.

On the other hand, a method of age determination by use of radioactive potassium (K) and one of its products, argon (Ar), is applicable directly to some sedimentary formations. The mineral glauconite is a silicate containing potassium and iron; it is formed by precipitation in granules in muddy or sandy seabottoms where there is little movement of water. Therefore, it occurs in some marine shales and in metamorphic rocks derived from these. The potassium–argon method can also be used on biotite, a black, potassium-bearing mica found in schists and igneous rocks.

Carbon is universally abundant in animals, plants, and of course in the organic compounds produced by them, as well as in carbon dioxide in the atmosphere and carbonic acid in water. A radioactive isotope, carbon 14, is produced by the impact of neutrons on carbon atoms in the atmosphere at high altitude, and being uniformly distributed, it occurs in a fixed proportion to ordinary carbon in organisms. When an animal or plant dies no more metabolic changes take place but the disintegration of C^{14} proceeds at a rate indicated by its half-life of 5560 years. Because the amount is not large to begin with, it becomes impossible to measure with sufficient accuracy after about 40,000 years, but this method has proved extremely valuable for determining the ages of archeological relics, or various levels in prehistoric human sites, and of the wood, bones, teeth, and other organic remains found there. But its use in paleontology is limited because of the much greater age of most fossils.

As a result of the studies so far made, a moderate number of dates that have no more than 3 to 10 percent probable error are available, but these do not include every period. By far the greater part of all geologic time, perhaps 4 billion years, fell in the Cryptozoic, or pre-Cambrian, eon, and subsequent to this was an age of some 600 million years during which nearly the whole fossil record accumulated. The name Cryptozoic, for pre-Cambrian time, refers to the fact that most evidence of the existence of life in that time is indirect, fossils are exceedingly rare, and of course the time of origin of life is wholly obscured. This is partly because of the vast amount of metamorphism, erosion, and disturbance of the Cryptozoic sedimentary rocks.

Age determinations in some of the pre-Cambrian rocks of Canada show 2200 to 2600 million years. There are dates of 1600 to 1700 million years in Labrador, and of 800 to 1100 million in the Adirondack mountains of New York. Kulp (1961) gives figures that are recent and reliable for many points in the Paleozoic, Mesozoic, and Cenozoic time scale. For the boundary between middle and late Cambrian the age is 530 m.y.b.p. (million years before present). There is not yet full agreement as to the position of the base of the Cambrian in relation to the rocks themselves, but provisionally an age of 600 m.y.b.p. is suggested. The base of the Ordovician is quite definite, at 500 million years. The Silurian is shown to be the shortest of the periods, about 20 million years, and the beginning of the Devonian is dated at 405 m.y.b.p. Some other significant dates are:

Beginning of Permian, 280 m.y.b.p.
Beginning of Upper Triassic, 200 m.y.b.p.
Beginning of Cretaceous, 135 m.y.b.p.
Beginning of Cenozoic (Paleocene), 63 m.y.b.p.

The average length of a geologic period is approximately 60 million years, but the range is from about 20 to 100 million years.

EVIDENCE FROM PALEONTOLOGY

As far as the biologist, and more specifically the comparative anatomist, is concerned, the major problems for him to solve in the study of fossils from earlier times are essentially the same as those in the study of evolution of recent animals, but because of the limited evidence available to him, he meets difficulties that are peculiar to paleontology and that could easily be surmounted or might not occur at all if he were studying recent animals. The first of these difficulties concerns the interpretation of structure and the "restoration" of extinct animals.

Fossils fall into several categories in regard to the nature of the evidence that they can provide. Very commonly fossils are not identifiable. That is, they may consist of fragments lacking any characters that would differentiate one kind of animal from another. It may be possible to tell that a fragment of a limb bone came from a mammal, and even from a large mammal or a small mammal, without being able to determine in any way whether the mammal was a carnivore, an ungulate, or something else. Then there are fossils which, being a little more complete, show one or two characteristics that permit identification, either by comparison with better specimens in a collection or by a person who is familiar with material of the same sort. This might, for example, be the case with a broken piece of a molar tooth that included a part of the occlusal surface or the end of a femur showing the processes and articular facets. Then there are many incomplete but recognizable specimens that one collects in ordinary field work, such as teeth, either alone or in series, complete but perhaps isolated bones, jaws bearing a few teeth, partial skulls, skulls in good condition, and so on. Occasionally, by chance, a minute fragment of an animal, if it happens to include a part bearing a distinctive characteristic, may easily serve for the recognition of a genus or species. Less commonly but still not rarely, the paleontologist finds excellent skeletal material in the field, associated or even entirely articulated, and is able to put it together and describe the entire skeleton of the animal concerned.

Difficulties met in the interpretation of the general run of fossils are fairly obvious. The paleontologist does not "restore an entire animal from a single bone" by any esoteric laws known only to him, but he can simply recognize an animal from its tooth or from its lower jaw or from a foot. This recognition implies merely that in knowing the kind of animal, he knows in a reasonably accurate way the characteristics of the rest of it, the parts that he does not have; but this is because he has seen other more complete material of the same or related kinds, or he knows from the work of other paleontologists what the material is like.

Suppose, then, that his specimen is reasonably complete, or if not complete is entirely recognizable, and he is able to picture the remainder of a skeleton from the part that he has: does this solve all his problems of interpretation? In recent years increasing attention has been paid to the restoration of parts that are not directly represented in fossils at all, such as muscles, connective tissue, blood vessels, nerves, and brain. The evidence on which such restorations are based is, of course, provided by the parts that are preserved, bones and other hard tissues, but the soft anatomy must be read in, so to speak, to the anatomy of the skeleton.

Muscles, for example, leave on the bones to which they were attached some indication of their presence in the form of smooth areas showing the location of fleshy origins and insertions, raised ridges along the edges of those attachments, projecting processes in the form of knobs, ridges, or points to which either muscles or the tendons of muscles were connected, and so on. By comparison with the skeletons of recent animals in which there is some resemblance to the fossils, it is possible to decipher these indications of muscular anatomy and construct a reasonably accurate picture of at least some of the muscles of the body in a fossil. Skulls and frequently other parts of the skeleton show openings, foramina, canals, or grooves that have been occupied by arteries, veins, or nerves, and it is usually possible to determine from the anatomy of recent animals which of these is represented in a particular case. Thus the nerve supply and blood supply to particular parts of the head in a well-preserved fossil skull can sometimes be pictured quite completely. In the same way, the cavity occupied by the brain may be filled with a cast composed of the matrix in which the fossil was enclosed, or if the cranial cavity is empty an artificial cast can be made that will show many of the characteristics of various parts of the brain. By such work as this Tilly Edinger built up almost single-handed the science of paleoneurology, which is now a large and complex branch of vertebrate paleontology.

Specimens are occasionally discovered in which there are natural casts of soft parts in the body cavity; for example, the stomach or intestine may be filled with a matrix of finer grain than that which surrounds the animal or which has filled other parts, and therefore a natural cast of these organs is produced even though the organs themselves are of course totally destroyed. Similarly, imprints of either bones or the soft parts such as skin or feathers of extinct animals may provide evidence of the shape and the nature of the surface of such parts even when the organ that produced them is long since gone. In this category are the footprints of extinct animals, which have rarely been associated directly with the skeletons of the creatures that made them, but specialists in fossil footprints are able today to make reasonably accurate restorations of the feet and sometimes the general proportions of the bodies of the animals that made such footprints.

But after all this, even with the best of fossil evidence, the most complete sort of material and adequate series of specimens of each kind, the paleontologist is left with little or no information concerning much of the anatomy of extinct animals about which he might wish to be informed. He must make the best of what he has. It is actually a great deal. But whether much or little, in any particular case he next is faced with problems of classification and evolutionary relationships. These, even if we disregard incompleteness of evidence, are somewhat different from such problems encountered in the study of recent animals. Only a classification based on morphology, and then only on that part of the morphology that is actually shown in the fossils, can serve the paleontologist, for he has no way of determining color of hair, distribution of skin glands, the reactions of certain proteins, and so on. Usually he has no information on life histories, methods of reproduction, the rate of growth, the age or sex of individual animals, or the stages of their ontogeny. This restricts the evidence to particular categories, but it does not render that kind of evidence any less reliable.

It must be remembered that classification is one of our methods of organizing the enormous quantity of information about nature into a form or pattern that can be handled in a reasonable way, preferably as simply as is consistent with the facts and in a way that expresses what we suppose to be a natural order in the world. The concept of evolution has given us an approach to such order in the plant and animal kingdoms, provided we can apply our understanding of it to this use. After a hundred years of evolutionary biology we are still relatively ignorant of the enormous variety and quantity of organisms. Our progress in classification has been very great in that time, as it has in the study of evolution, but usually it is necessary that some classification be made before evolutionary study, or any other kind of study, is possible. This is because we must know the characteristics of organisms to some degree before relationships can be appreciated.

In practice, classification is developed by adding bit by bit to the structure already raised, as new species, new genera, and other categories, living or fossil, are discovered or for the first time understood. This process involves not only describing and naming but also, in a sense, elbowing aside a crowd of other species and genera to make room for those that are to be fitted in. Frequently the new evidence requires a resorting and rearrangement of previous classifications of animals already known. This is likely to go on throughout the foreseeable future, even in the study of organisms that are living today, but a paleontologist also has a virtually infinite treasury of little-known or yet-unknown forms of life. It is likely that his new discoveries will continue at a constant if not an increasing rate for many future generations, even when our knowledge of recent animals has begun to reach the point of diminishing returns.

The fact that animals are fossil does not mean that they require, or can even be conceivably fitted into, a separate or different type of classification from that of recent animals. Instead, they supplement the knowledge of existing animals, by carrying back into the past the ancestral lines as well as lines that appeared and then became extinct. Fossils show us in this way a four-dimensional world of life.

The special difficulties of the paleontologist as regards classification are several. One is that the fossil record is biased, in the sense that preservation of fossils is selective. The animals and plants preserved are those that have parts which can endure burial and remain long enough to form part of a compact and solid stratum or, if not solid, one in which bones or teeth at least can resist decomposition. It is improbable that such animals as earthworms, caterpillars, maggots, or jellyfishes will fossilize. This is not quite impossible (see Fig. 34, *Aysheaia*); indeed, a few impressions of jellyfish in soft sea mud are known to have been fossilized, but it is highly unlikely that such a group of animals could produce any evolutionary record of itself.

In a more limited sense, even among the vertebrates fossilization is highly selective, for much depends upon the nature of the sediment and ecological situation in which it is formed. Limestone deposited in the sea contains few or no fossils of land birds or terrestrial mammals. The sand deposited in windblown desert dunes and later consolidated into sandstone is unlikely to contain whales or fish. Stream-deposited sediments are similarly selective in regard to the animals that are commonly buried in them. The result of all this is that in our knowledge of fossils there is

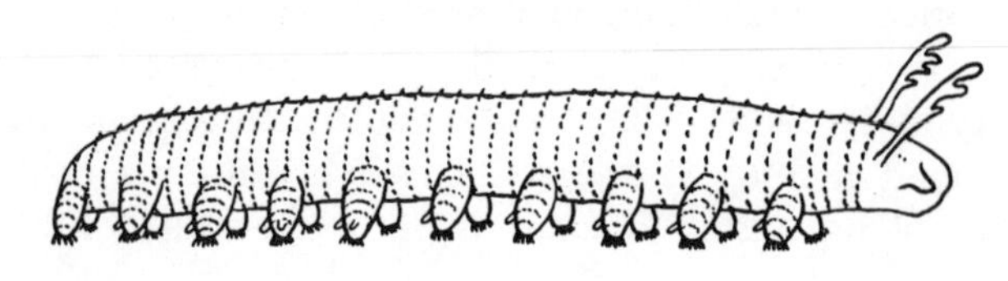

Figure 34. Reconstruction of a middle Cambrian onychophoroid, *Aysheaia pedunculata.*

strong emphasis on particular groups and particular environments and a virtual absence of information about other groups or other environments.

Another potentially serious problem confronting the paleontologist is that the more complete his information becomes in regard to the connections of various lines of descent with one another in the past, the more completely the established categories blend into one another and the greater the difficulty of defining each of them will be, until finally, assuming a full knowledge of evolutionary relationships in a group of animals, there should be no point at which one genus can be distinguished from another, one family from another, or one order from another. All are connected, if their lines are followed down to the points where they join in common ancestry in the remote past. It is true that one can still distinguish a branch of such a tree from other branches, but only after the branches themselves have separated, for at the point where they meet the characteristics on which separations have been based are blended together. Therefore, the fullest possible knowledge of phylogeny would result in the complete obliteration of a system of taxonomy such as we have today. This may not be a calamity but rather a challenge to devise some other workable scheme to fit a continuous but branching type of variation such as we see in a four-dimensional picture of the evolution of organisms.

Nevertheless we are not likely in the future to meet this problem on a very large scale. On one hand, there will be the selectivity and imperfection of the fossil record, which makes the complete continuity of an evolutionary picture impossible. On the other hand, there is the difficulty, indeed impossibility, of finding all the fossils that have been preserved, because the only fossils available are those that have been brought to the surface by erosion or are encountered in chance diggings, roadcuts, mines, wells, and other artificial surfaces created by man. Furthermore, classification will, as has been implied, work in much the same manner as it does today as long as it refers to the differences and resemblances among the branches of the evolutionary tree that have been established. The difficulty comes in those relatively small segments of the branches that lie close to the point of divergence of one branch from another.

REFERENCES

Kulp, J. L. 1961. Geologic time scale. *Science 133:* 1105–1114.

Moore, R. C. 1958. *Introduction to Historical Geology,* 2nd ed. McGraw-Hill, New York.

13 · ORIGIN OF LIFE

SETTING THE STAGE

Throughout the known universe, as shown by spectroscopy, the analysis of meteorites, and the recent probes into space, the elements and many of the compounds known on earth are widely distributed. We meet no processes or forms of energy that are inconsistent with our understanding of the physics and chemistry of the earth and sun. In the galaxy or "island universe" of the Milky Way, of which our solar system is a small part, it is thought that some, perhaps many, of the millions of stars are suns of other solar systems; that is, they may be attended by planets as satellites.

Among these in turn a small proportion may fulfill certain conditions that would allow them to be inhabited by some kind of life (Shapley, 1963): (1) The sun for such a planet must not fluctuate noticeably in its production of light and heat but maintain a steady output through at least some billions of years. (2) The orbit of the planet must be nearly circular; otherwise, energy received from its sun would vary too far to permit stable conditions. (3) Either liquid water or (improbably) some other versatile solvent must be present as a medium for the action and evolution of highly complex chemical systems. (4) Obviously such systems must not meet severely injurious or inhibitory materials or processes.

Considering that the number of galaxies is in the hundreds of millions, Shapley infers that habitable planets are abundant, even though unthinkably remote from one another, and that great numbers of them "must also have experienced the natural evolutionary processes" that lead to the differentiation of organisms.

Thus far, probably, we may agree. These premises, however, should not lead us to conclude that life has taken the same forms on other planets that it did on the earth. There is no reason whatever to suppose that evolution elsewhere produced animals, plants, chlorophyl, flowers, trees, insects, fishes, reptiles, or man. Skeletons, muscles, and nerves need not have arisen inevitably, or at all. There are other kinds of action—the flow of liquids, turgor, waves, and ciliary vibration—and a virtually infinite variety of shapes and structures.

Evolution is a process that works in living things as they happen to be, adapting and maintaining them in the conditions that are given. It is not a path to a foreordained future; it does not anticipate; it has no "goal": not man, nor intelligence, nor any other. Conceivably intelligence might evolve as one of many possible adaptations of behavior in more than one planet, but this is not predictable. Considering the enormous number of heritable traits that were established (and later changed) in adaptation to special circumstances in the long line of human ancestry on earth, man must be one of the most improbable productions of this universe. The stage of life that is open to our investigation is, so far, only the earth, and we are most unlikely to find another.

The manner in which this stage was set is not yet fully understood, but much progress has been made in recent years under the well-supported assumption that planetary systems, stars, and indeed galaxies themselves are not all of the same age but severally exhibit phases of early development and gradual transformation. Urey's book (1952) on the origin of the planets is especially important in regard to the source of the earth and the primitive conditions of the lithosphere, hydrosphere, and atmosphere.

In the almost complete vacuum of space are many nonluminous "clouds" of rarefied gas (mostly hydrogen) and dust [fine crystals of ice, ammonia (NH_3), methane (CH_4), particles of metals, silicates, carbonates, and other compounds] so thinly scattered that space is virtually empty and the temperature close to absolute zero. Some of these "clouds" are less than a light-year in diameter (a light-year is about 5.86 trillion miles), but others are enormously greater.

It is suggested that the total mass of one of the smaller globular clouds is comparable to that of our solar system and that the latter developed by condensation of such a cloud, because of the mutual gravitational attraction of the materials in it. As a denser core began to form the process accelerated, gradually drawing a large proportion of the gas and dust to the center and imparting to it a rotational movement, which in turn tended to localize the remaining matter in a single plane, or disk. With the enlargement of the core, its accumulation of perhaps 90 percent of the substance of the original globule, and its resulting great density, a point was reached at which pressure and the radioactivity of some included elements raised the temperature to thousands of degrees centigrade. Presently hydrogen began to transform into helium. Atomic disruption and fusion became predominant over chemical change. Ordinary reactions were overwhelmed in a tremendous outpouring of radiation as the nucleus of the cloud became a sun with an internal temperature of millions of degrees. Hydrogen and helium remained, however, the most abundant elements, far greater in mass than all the others in spite of their low atomic weight.

Meanwhile the outlying disk, which by reason of its rotation was not drawn to the center, underwent locally similar episodes of concentration, forming solid or gaseous bodies with gravitational forces in proportion to their masses, and these swept up in their orbital journeys the remaining gases and much of the "dust" left over from the cloud. As satellites of the sun they received its heat and light, the effect upon them being inversely proportional to their distance. Continuing to grow by accumulation of matter from space, they, too, might reach a mass that in condensing produced internal heat great enough to melt most solids. This made possible a selective shifting of materials by weight. The relatively heavy compounds of iron and nickel sank toward the center, leaving those that were lighter (silicates and carbonates) in the more superficial zones, forming (at least in the minor planets) a solid crust. In the earth this process continues, as local areas of heating and melting affect the crust, and as molten rock from time to time presses into new cracks, faults, or zones of weakness.

Most of the original gases included in the condensing earth were retained as a primitive atmosphere of ammonia, methane, water vapor, and hydrogen, the latter escaping most readily. The lightest gases, except what were bound chemically with other elements, accumulated in the sun. Much more slowly methane and ammonia escaped the earth's gravitation but were largely retained by the giant planets, Jupiter and Saturn, where they form important components of a dense atmosphere. It has been shown with high probability that the original atmosphere of the earth did not contain free oxygen, because oxygen was not released from compounds of carbon, hydrogen, silicon, and other elements in any significant quantity until photosynthesis began.

Although the number of publications bearing on the origin of life is now very large, the student will perhaps be served best by reading Oparin (1962); this incorporates recent work and is not, therefore, merely a restatement of that author's earlier studies. He, more thoroughly than any other, has discussed the probable stages in the beginning and earliest evolution of life. Miller (1953) and numerous workers in the last ten years have brought forth direct evidence on the early synthesis of organic compounds that took place in the ages before life, when reduction rather than oxidation was the primary chemical source of energy for organic reactions. Before considering these findings it would be well to make some statement of the meaning of life, organism, and organic evolution in relation to this problem.

NATURE OF LIFE

We can expect more and more of the fundamental questions of biology to be answered, at an ever-increasing rate, as long as our approach is like that of most scientific work of the past hundred years. Its basis is that observation and, where pertinent, experiment are the sources of our knowledge of the natural world. Observations, made independently and yielding similar results, may be considered correct as far as they go. An experiment is a procedure in which a process already observed is altered in such a way that the result of the alteration will answer a question that could not be answered by observation alone. The immense store of information that we now have on the subject of life, organisms, and evolution obtained in these ways enables us to predict the results of future observations with complete success in many cases and to put our knowledge into a consistent, noncontradictory system. Its fruitfulness in understanding is the most impressive of all human achievements.

To a biologist the words "life" and "organism" mean little apart from each other, and evolution is as much a characteristic of life as reproduction or metabolism. An organism is an "open" system that takes matter and energy, acts upon them, changes, and releases them, maintaining its own identity by doing so. Life is the name for this activity. Along with the continued flow of interacting materials there is a coordination of the processes in such a way as to preserve a relationship with the environment, a kind of fitness of one to the other. This takes two forms: a survival of the organism as such, and survival of some or all of its components by means of replication and reproduction. The ability of organisms to regulate, in some measure, their own processes, to control in part their own survival, does not occur in other kinds of matter, including machines. Adaptiveness of a machine was built into it by its maker, but even the most complex devices of modern technology do not compare in this respect, or in complexity of function, with living organisms.

Because organisms grow and multiply, and because individual survival and reproduction are in part subject to heritable differences among individuals, there is among all populations of living things a certain adaptability to the environment by means of differences in reproductive rate and survival. This process, which gradually changes the genetic composition of successive generations in a species, is natural selection. If an environment remained constant, selection would act only to limit maladaptive changes in a population. But no environment is unchanging; not only do its physical factors vary, at least slightly, but the equally important biological factors, represented by interaction among organisms of the same or different kinds, also vary. Thus any population is modified in the course of time by selective adjustments of its heritable characters, slight though this may be. This is biological evolution.

CONDITIONS FOR THE ORIGIN OF LIFE

Before life began there was evolution of another kind, not involving precisely reproduction, heredity, and natural selection but a somewhat comparable accumulation and change of complex organic materials. The process must have been more nearly random, less definite in its limits, and far slower in its cumulative effects. Certain characteristics of life (growth, division, colloidal properties, oxidation, and so on) can, of course, be found among materials that are not living. In fact, it is impossible that the many distinctive properties of even the simplest organisms arose at one step, simultaneously. Therefore the "origin of life" was not an event of a particular moment or day but a gradual, prolonged episode of indeterminable time.

When the earth was established, some 4 to 5 billion years ago, with a solid crust and an atmosphere of hydrogen, methane, ammonia, nitrogen, carbon dioxide, and water vapor, it is thought that there was relatively little liquid water on the surface, perhaps one tenth of the amount present today. The remainder was bound up in the chemical composition of rocks and, later, gradually released from them. Nevertheless, the geological work of erosion by weathering, solution, and transportation must already have begun, and exerted throughout all subsequent time an influence on the shape and composition of the earth's surface. Streams running into the basins of the ocean bore with them in solution many compounds taken directly from the crust or dissolved, on the way, from the rocks and atmosphere, thus gradually accumulating in the young sea a complex mixture of solutions. It is likely, however, that at least half of geological time elapsed before it became possible for protoplasm to develop in this solution.

Considerable light has now been shed on the way in which this could be brought about, by the ingenious experiments of Miller (1953) and others who followed him. A protein molecule is an extraordinarily complicated structure of amino acids, the small molecules of which can be combined in an infinitely diverse pattern of chains and linked sequences, but the number of kinds of amino acids available is only about twenty. The precise pattern taken in a particular kind of protein is unique to it. One question that had to be answered, therefore, was the means by which, in the absence of any previous life, the amino acids themselves appeared. Using the components of the primitive atmosphere in a closed chamber and providing silent discharges of electricity through the chamber, Miller demonstrated that many of the common amino acids were actually produced from a mixture of ammonia, methane, water vapor, and hydrogen. All these are known to have been present and abundant in the early atmosphere, through which, at least occasionally, electrical discharges of lightning must have passed. By adding to the experimental mixture carbon dioxide, carbon monoxide, and nitrogen, Philip Abelson in 1957 was able to produce more than twenty amino acids, including all those commonly found in living matter, plus some proteins.

It is therefore not surprising, but rather inevitable, that organic matter of high complexity should appear in the primitive ocean. But this was not life, for protoplasm and organisms still lay far in the future. There must have been a great number of random chemical reactions taking place in and among the accumulations of organic and inorganic matter in solution, some of which resulted in forming complicated polymers of amino acids and numerous locally concentrated solutions of different physical and chemical natures. In fact, because no organisms existed yet, no destruction and consumption of these solutions by bacteria or protozoa could take place, and it is probable therefore that they reached a higher concentration and approached more closely to the colloidal state than would have been possible in any ocean during the last 2 billion years. To express it in another way, one of the necessary circumstances for the origin of life was that nothing living then existed, and the converse is that *since* the origin of life, it is unlikely that life could originate again, independently, because organisms already in existence would destroy such concentrations of organic substances long before they reached a protoplasmic stage.

It has been suggested by Oparin and others that a necessary mechanical step toward the formation of the first organisms able to sustain themselves could be taken by development of separated droplets, as in an emulsion, made by the mechanical mixing of one complex solution with another. Such are called coacervate droplets. They require, of course, that two solutions should not be completely miscible in one another and that the droplets formed by mixing them have some kind of surface membrane that at least partially isolates the material inside from that outside. Many years of experimental work on such coacervate mixtures have shown that the droplets frequently have the property of adsorbing from their surroundings particular dissolved substances, even when these occur only in exceedingly small concentration. The droplets also may or may not develop osmotic relationships with the medium in which they are enclosed. Given diverse mixtures and solutions, and any sort of agitation, such as breaking of waves or a current tumbling downhill, the formation of coacervate droplets must have occurred inevitably, but the chemical and physical episodes could not have occurred in any sort of order, nor for a long time have developed any self-regulation or continuity.

It is not necessary to imagine that by some rare accident, a pure chance occurrence once in a billion years, a molecule comparable to a virus and able to reproduce itself from then on, appeared in the primeval ocean. Indeed this is more than improbable, for the self-replication of viruses and nucleic acids requires the presence of a highly adapted medium of enzymes and proteins, the functions of which are specifically to bring about such replication. Thus there must have been in the earliest proto-organisms chainlike reactions that took from the environment particular materials and in a cyclic or repetitious manner transformed them in ways that promoted the continuity of the droplet in which the reactions occurred. Such cyclic reactions could have occurred without necessarily leading to the maintenance of a population of preorganisms, but occasionally, among the reactions in such droplets, some may have taken place that furthered one or another of the lifelike properties of the system.

For example, the survival of the droplet would depend on stability of its surface film. Yet too firm or too thick a film could prevent the intake and expulsion of materials necessary for maintaining the droplet. Droplets in which the mass of material being taken in slightly exceeded over a period of time that which was released would of course grow, but whether as a result of growth they became unstable and destroyed themselves by falling apart, or whether the surface film kept its own identity and maintained the fragments as individual droplets themselves, would depend upon the nature of the film and its ability to maintain itself under mechanical strain.

Another essential step must have been the occurrence of reactions that in turn stimulated or accelerated other essential reactions, thereby increasing the efficiency of what we might call the metabolism of the droplet. Countless chemical changes are possible at various rates among mixtures of compounds such as we are picturing, but if self-maintenance depends upon the effectiveness of some of these reactions, then any accelerating mechanism serving the function of a catalyst could facilitate enormously the survival of the system. We can imagine that catalysis of a few, and eventually many, and even all of the reactions occurring in a particular kind of droplet was developed step by step through differential survival, in other words, by a primitive kind of natural selection.

For survival, however, not merely speed but coordination and equilibrium of the reactions is essential if the system as a whole is to keep its stability. This would be lost just as easily by excessive rates of reaction in one kind of metabolic process as by sluggishness in others. Therefore the increasing integration and linkage of reactions would have to go hand in hand with improvement in organization of each kind of process. Only efficient systems survived, and in the long run only those capable of reproduction could progress beyond the inanimate stage of coacervate droplets. Prerequisite for reproduction is, of course, the property of growth, and prerequisite for that is the organization of an open, dynamic, but stable system, one in which energy from outside is continually used in the reactions concerned with self-maintenance and self-renewal internally, and in which the total accumulation of matter and energy in a given time slightly exceeds their depletion.

At this point we might say that the kinds of droplets in which such processes developed had a history. Oparin emphasizes the enormous length of that history, probably half of all geologic time, and its complexity. The precise nature of the common metabolic reaction sequences in one kind of droplet depended upon their past. The same effects could in many cases be accomplished by any of several different reaction pathways, but those that were, so to speak, chosen at each critical stage in the history of these proto-organisms would, of course, determine the foundation on which further advance was made.

BEGINNING OF BIOLOGICAL EVOLUTION

When the division of droplets became a regular, recurring event associated with their metabolism and growth, they could cross the dim line dividing the lifeless from the living. This does not imply any particular novelty in the process of living, but the introduction of the biological phenomenon of differential survival. Survival of a given lineage of dividing units was continued and sustained merely by its success, whatever the circumstance to which this was due, while another lineage under similar conditions might fail and disappear. Successful evolution from that time on was that which continued, and unsuccessful evolution was that which ended with the extinction of the organisms concerned. No other concept of values, of "fit" or "unfit" organisms, of "better" or "worse," can well be applied than this, and then only when referring to specific circumstances.

The introduction of natural selection in a biological sense (that is, with variation leading to differences in rates of survival and reproduction) must have resulted in a decrease of randomness in the activities of the organic units populating the primeval soup and in an increase in the rate of progressive change in the efficiency of these units. Organisms, self-sustaining, reproducing, and subject to natural selection, have probably existed for more than 2 billion years. At least half of that must have been occupied in the gradual improvement of cellular organization, perhaps to a stage no further than development of the simplest colonies of cells or the origin of the earliest Metazoa, or sponges, or filamentous algae and fungi.

No doubt within this time a major event was the establishment of a nucleus as a regulatory center for some activities of cells, of the elaborate chain molecules of deoxyribonucleic acid as the means of replication of the protein components of successive generations of cells, and of ribonucleic acid as an intermediary in this fast, incredibly complex process.

Probably also in this early time came the reduction of the richly organic but nonliving "soup" of the primeval ocean by means of the activities of organisms. Although it was the matrix in which life originated, it could not survive the highly efficient digestive and metabolic processes to which it was now subject.

Likewise in that remote era came the earliest experiments in photosynthesis by green algae, which had the vast importance of releasing free oxygen into the water and atmosphere of the earth, therefore gradually making this element available for energy-producing reactions of oxidation. Although it seems clear that the metabolism of the earliest organisms must have been anaerobic, and although some primitively anaerobic reactions remain as a heritage in recent organisms of all kinds, nevertheless this is overlain in most cases by an accumulation of metabolic processes that require oxygen. Intake of oxygen from the surrounding medium and its transportation and uses within the body come into a poorly definable category of respiration. But the other side of the same picture is that photosynthesis evolved as a device for producing food within cells—carbohydrates in excess of the organisms' demand at any given moment. It was one of several autotrophic ("self-feeding") processes, and by far the most successful, that were developed in primitive organisms which had been heterotrophic, dependent originally on taking in complex organic compounds from their environment.

Both the carbohydrate accumulation and the release of oxygen by photosynthesis led to a vast transformation of the environment and its adaptive potentialities for nonphotosynthetic organisms. Now saprophytes (as fungi) and active heterotrophic organisms (animals), both requiring oxygen in their metabolism, could evolve, their primary food supply coming from photosynthetic plants. The continual interaction among all these provided innumerable adaptive pressures, leading to differentiation of groups that were specially equipped for various kinds of reproduction, life in surface waters or deeps, in sea or fresh water, on land or even in the air, to herbivores on which in turn fed carnivores, both of them providing sustenance for parasites.

At an ever-increasing rate, the water and then the land and air have been occupied by ever more diverse forms of life. At a time between 500 and 600 million years ago, in the Cambrian, there came, apparently, a relatively sudden development of animals in which hard materials, especially compounds of calcium, were deposited as supporting or protective structures. Thus the mollusks, brachiopods, bryozoans, corals, echinoderms, and arthropods, among others, began to leave fossils of their hard parts in the marine sediment of that period. This was the beginning of the fossil record that continues to the present time. A few fossils having the form of impressions of soft-bodied organisms or carbonaceous films left by aquatic plants occur in pre-Cambrian rocks, but they are nowhere abundant, and they lack either the continuity or the structural detail to give us much information about the history of pre-Cambrian life.

REFERENCES

Miller, S. L. 1953. Production of amino acids under possible primitive earth conditions. *Science 117:* 528–529.

Oparin, A. I. 1962. *Life: Its Nature, Origin, and Development.* Academic Press, New York.

Shapley, H. 1963. *The View from a Distant Star.* Basic Books, New York.

Urey, H. C. 1952. *The Planets, Their Origin and Development.* Yale Univ. Press, New Haven.

14 · RECOGNIZING EVOLUTIONARY RELATIONSHIPS

The problem of determining relationships between ancestors and descendants and picturing as accurately as possible the genetic connections among a diverse group of animals or plants is faced continually by the paleontologist, but it is not so frequently a concern of other students of evolution. It requires that we recognize changes taking place in the evolution of characters from a primitive to a specialized condition, see the stages through which such changes were made, and infer the correct sequence of these stages.

It has been estimated (Romer, 1949) that not more than 1 percent of the genera of mid-Mesozoic tetrapods contained direct ancestors of any animals living today. The closer any past period is to the present in time, the greater is the proportion of such connections, but even when relatively close there remains the problem of distinguishing among the known animals of the earlier period those that could, and others that could not, have been ancestral to particular kinds living now. Of a somewhat different nature is the question of deciding the evolutionary relationships of a group of organisms for which there is little or no fossil record.

Certain principles are followed, but usually without explicit statement, by those who undertake this kind of work. Naturally the first step in such a study must be that of determining resemblances or differences in structure among the animals that are available for comparison. In certain cases, as when the characters are simple and quantitative in nature, resemblances may be misleading because such characters might be attained independently in different lines and their evolution may be subject to reversal in direction. (The idea that evolution is irreversible is not true for the simpler kinds of change.) In fact, quantitative characters by themselves cannot give any indication of direction of evolution, but a comparison of many characters, including some that involve complex mechanisms or configurations of parts, may be accepted with more confidence as showing the direction of descent as well as closeness of relationship.

In comparing animals of the present time with those in a fossil fauna that is recent enough to be roughly similar or in comparing fossil faunas with others, earlier or later, the problem of recognizing lines of ancestry and descent is often simplified, in that it becomes little more than a question of identification. If, among the mammals of the late Pliocene, say 4 or 5 million years ago, we seek those that could be ancestral to species of horses, rhinoceroses, or elephants living today, the field of choice is narrowed to only a few alternatives merely by recognizing the groups. In each group it is probable that some may be eliminated from further consideration by having diverged structurally away from a combination of characteristics that could most readily be linked with a descendant of the present time. Some differences that can be attributed to evolution would of course be expected in a descendant, but over a short time, geologically, these are likely to consist of minor changes of proportion and slight modifications of parts in a direction away from those which we see represented in the earliest members of the stock.

It is always possible that precisely ancestral species did not occur in the area from which the fossils are known or through the accident of sampling are not present in the collection. Nevertheless, geography is an important consideration when a fairly short time is involved. It is more likely that land mammals living today in Africa are descended from African mammals of the Pliocene than that they have been derived from Pliocene faunas of Asia, the Americas, or Australia, although in principle the latter might not be totally impossible. Some groups of organisms disperse far more rapidly than others. Although the Old World monkeys and apes (the primates most nearly related to man) have lived in Africa, Asia, and Europe for many millions of years, as shown by their fossils, there is no evidence that any of them ever reached the western hemisphere, nor did man do so until he crossed Bering Strait some 20,000 years ago when it was probably an isthmus.

General resemblance and recognition of a group of similar species are not enough by themselves to assure the connection of ancestor with descendant. Even if the fossil record were complete between two selected stages of evolution, the lineages would not necessarily be clear, unless the animals in a phyletic line were conspicuously different from those in any related line and there were little or no branching.

Obviously the evolutionist includes some other principles in his thinking. One must be a reasonable basis for distinguishing primitive characters from those that are advanced; the time sequence by itself is not enough, for at any time there are animals with primitive, and others with advanced, characters living as contemporaries. It is true that we cannot separate characters from the animals in which they occur, nor can we forget that adaptation means the ability of populations (rather than characters or individuals) to perpetuate themselves. Yet if we consider any feature of a phyletic line that is, over a long span of time, subject to change, the condition represented at an early stage of such change must be called primitive in comparing it with the different condition of the same (homologous) feature that appears later. This is all that the terms "primitive" and "advanced" mean. They are purely relative, to contrast ancestral organisms with those that have changed during descent.

The words "generalized" and "specialized" differ in almost the same way but have a certain additional meaning which can be illustrated by an example, using some familiar features of teleost fishes. If we select a pair of contrasting characters in the teleosts, such as the location of pelvic fins in either the abdominal or the thoracic position, we find that a great many teleosts fall into each category. This is true among both the living and the fossil kinds that are known. What basis do we have then for deciding which of these alternatives is more generalized? Going beyond the limits of the teleosts themselves and considering Holostei, Chondrostei, and all the other bony fishes (Osteichthyes) that are known, it is seen that the occurrence of abdominal pelvic fins is almost universal, whereas the thoracic position occurs rarely. On this basis, not considering other characters at the same time, it seems highly probable that the teleosts in which pelvic fins are abdominal must have inherited this characteristic from earlier fishes in which it is known to have been the normal condition, rather than that the thoracic position is primitive, for in the latter case we could not point to an ancestral stock, and moreover we would have left the abdominally placed pelvics of teleosts unexplained.

Judgments thus derived represent varying degrees of probability, usually high. When one such conclusion is coupled with a similar interpretation of several other characters in the same group of animals, these reinforce each other to the point of virtual certainty. We note, for example, that an air bladder with a duct is characteristic of almost all the teleosts in which the pelvic fins are abdominal and also occurs in other groups of bony fishes, whereas the duct is absent in those teleosts that have pelvic fins in the thoracic position. This gives us quite complete assurance that the ductless air bladder and the thoracic pelvic fins were derived from those with duct and with abdominal position, respectively. This opinion is further supported by observing other correlated characters, such as the spiny rays in the fins of many teleosts, which are seen to be specialized in contrast with the soft rays found in the same fins of other fishes, and the protractile apparatus of the upper jaw, which contrasts with the fixed maxillary and premaxillary bones that are features of the earlier groups of bony fishes. There is no doubt in the minds of those who have considered these relationships that we can recognize *generalized* teleosts by a certain combination of characters; a generalized character is one that can distinguish an early group, out of which diverse evolutionary lines may have come. It is, for any of these lines, also *primitive.*

In each of these instances we have relied upon characters that are widespread among groups other than, but related to, those being considered, because such widespread characteristics could most easily serve as a source of the ancestral heritage of the group in question and by the same token are most likely primitive. But if such information is lacking, it may still be possible on other grounds to decide which of two alternative characters is primitive and which is advanced. One of these grounds is simply the probability, based on the characters themselves, that evolution went in one direction rather than the opposite, between two contrasting conditions.

To take an extreme example, it seems far more probable that a complex structure such as a leg or an eye could be reduced and gradually lost in adaptation to special conditions than that the opposite should occur, and it appear out of nothing. Thus it is much more likely that snakes were derived from reptiles possessing legs than that such reptiles were derived from snakes. This is, of course, reinforced by the earlier consideration that most other animals related to snakes (the great majority of amphibians, reptiles, mammals, and birds) have legs. It is more reasonable to suppose that the few kinds of blind cave fishes and salamanders have become eyeless from an eyed ancestry than that all normal fishes and salamanders evolved from blind cave dwellers. It is not necessary to bring into the argument the genetic complexity of such organs, as has sometimes been done, for although it is true that their long history of adaptation involved a tremendous number of genetic changes acted upon by selection, nevertheless to eliminate an elaborate genotype as a means of adaptation to new, special conditions does not require going through the same history in reverse. The growth rate of a complex structure may be under relatively simple genetic control, and variations in this rate may, under selection, bring about an atrophy of the organ, overriding the complicated genetic pattern established during its evolution. Indeed, there is reason to think that a long evolutionary history, involving numerous adaptive modifications, does not require the accumulation of a vast number of items in the genetic code but rather the conversion and partial reuse or readjustment of genetic material that was present in the ancestry of the animal.

The utility of quantitative characteristics in studies of evolution calls for comment here. Innumerable studies have been made of such characters for purposes of classification. In snakes, for example, the number of scale rows, ventral scutes, labial plates, precaudal and caudal vertebrae, and so on, can be counted. In all these features the counts in a species usually vary somewhat but cluster around a certain average for each characteristic. Other related species may have the same, nearly the same, or different averages, and some overlapping often occurs between the counts for a given character in two or more species. Is it possible, when such figures are obtained for a large number of species and genera in a taxonomic group, to draw any conclusion regarding the trend of evolution, or the direction in which the number of vertebrae, scutes, or scale rows has gone?

Similarity or difference is demonstrated in these figures, and they may therefore be of some use in taxonomy, but their application to evolutionary questions does not follow, for we cannot, on the basis of the figures alone, make a choice as to what is primitive and therefore cannot determine the direction of change, *unless* there is other evidence of this, which can be correlated with the figures. Frequently that evidence is available, and as a result there can be general agreement on the evolutionary meaning of a certain numerical characteristic. For instance, among birds it has been found that the number of primary feathers in the wing differs in various orders, but that the smallest number (nine) is a specialized condition, reduced from a primitive count of eleven or more. In teleost fishes, among the spiny-rayed kinds, it is agreed that the advanced families have three spiny rays in the anal fin, although among other teleosts the number may vary from none to seven. Such conclusions obviously require the study of other characters of a different nature, such as the morphology of the skull, and often some help from the fossil record.

To other quantitative characters, which are not expressed in simple whole units but in more complex measurements, as, for example, precipitin reactions between sera of different kinds of animals, the same rule applies. It is often supposed that the similarity and difference in the blood proteins that are indicated by these figures actually indicate the relationship, in evolutionary terms, that exists between the species concerned. In fact, however, the figures only indicate similarity and difference of proteins from a particular body fluid. They show nothing of the direction or rate of evolution, and cannot, by themselves, indicate what is the primitive condition or the extent to which the picture is affected by parallelism, convergence, reversal, and delay or acceleration, all of which may affect quantitative evolution. It is sometimes argued that if the figures for many species tend to cluster, while a few stand at more or less remote distances, it is safe to assume that the ancestral stock was like the clustered figures. This is by no means the case, unless we have other grounds for recognizing the source of the group of animals, for there is no intrinsic reason a group of recent species should have characters that cluster around those of an ancestral stock; frequently the majority of species in a group represent a recent, successful radiation, whereas a more distantly related species resembles the common ancestor, and this may be expected to show in quantitative characters. In summary, then, numerical characters of species and genera are of some utility in distinguishing taxonomic divisions within a group (although wholly meaningless if lumped to make an "average" for a family or higher category!), but they cannot, by themselves, be used to make out evolutionary relationships as such.

On the other hand, configurations, or so-called qualitative characters, not amenable to simple quantitative description offer the clearest evidence of both taxonomic and phylogenetic relationships. They are complexes or aggregations of unit characters that are adaptively related to one another in an animal, so that their evolution proceeds in an integrated pattern. Such, for example, are the peculiar shapes and struc-

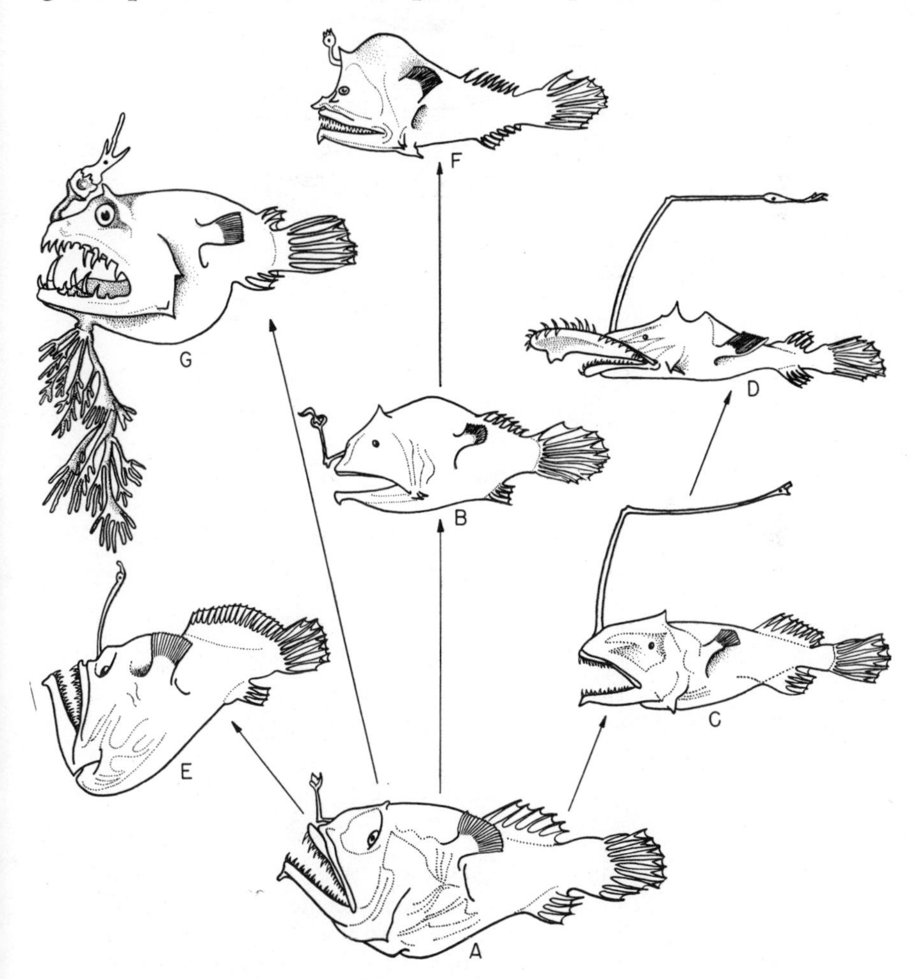

Figure 35. Adaptive radiation of body form and "lure" of deep-sea angler fishes (Ceratioids): A, B, C, *Dolopichthys;* D, *Lasiognathus;* E, *Melanocetus;* F, *Lophodolus;* G, *Linophryne.*

tural features of the genitalia of male insects in several different orders. Another case is the form of the skull in a series of bony fishes belonging to, let us say, an order or a sequence of fossil families and genera. The configuration of such skulls carries evidence of changes in adaptive uses of the head, mouth, eyes, gills, jaws, and other characters of the animals. Yet it must continue to evolve as a functional and successfully adapted unit in all stages of its history. Using such characters as this, one is able to make reasonable judgments concerning the direction in which changes have gone, the condition that was primitive, and the amount and kind of divergence from that, with or without a record of the time sequence (Fig. 35).

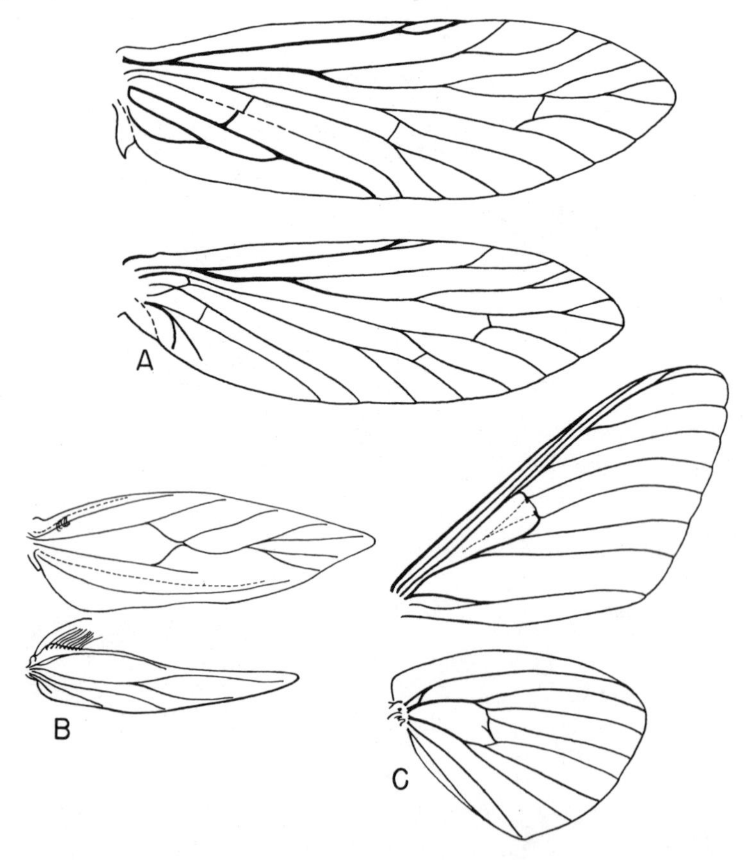

Figure 36. Wing venation of Lepidoptera: (A) *Mnemonica,* a primitive moth; (B) *Obrussa ochrefasciella,* a minute leaf miner; (C) *Citheronia regalis,* the regal moth. (Redrawn from J. H. Comstock, *An Introduction to Entomology.* Copyright 1920 by Comstock Publishing Company. Copyright 1924 by J. H. Comstock. Copyright 1933, 1936, 1940 by Comstock Publishing Company, Inc. Used by permission of Cornell University Press.)

In studying the wings of insects, Comstock, Needham, and others were able to infer the primitive pattern of venation from which that of existing orders of flying insects could have been derived (see, for example, Comstock, 1950, and, for a different opinion, Edmunds, 1954). This agrees with the arrangement of veins and particularly of the antecedent tracheae in the developing wings of some cockroaches, an order that has changed little since the Pennsylvanian. The pattern is also much like that of both fore and hind wings of the most primitive families of Lepidoptera (suborder Jugatae, Fig. 36A), the most primitive Diptera, Megaloptera (Fig. 37), and various representatives of several other orders of existing insects. Unmistakable evidence of modification from the primitive condition can be seen as the venation of fore and hind wings becomes dissimilar in

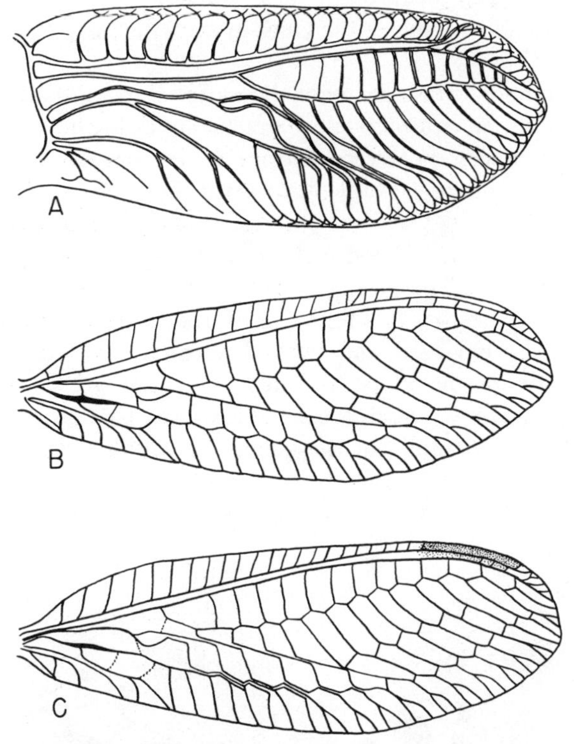

Figure 37. Venation of the fore wing of the golden-eyed fly, *Chrysopa nigricornis:* (A) pattern of tracheae in the wing of the pupa; (B) venation of the adult, showing the previous pattern modified by attachment of some of the veins to each other; (C) same as (B) but with the veins separated to indicate homologies. (Redrawn from J. H. Comstock, *An Introduction to Entomology.* Copyright 1920 by Comstock Publishing Company. Copyright 1924 by J. H. Comstock. Copyright 1933, 1936, 1940 by Comstock Publishing Company, Inc. Used by permission of Cornell University Press.)

advanced members of these orders and as the pattern is adaptively changed in either or both pairs of wings.

Certain veins present in the primitive wing may fail to develop, resulting in a simpler pattern (Fig. 36B). Loss or atrophy of veins here and there occurs frequently, especially in insects of small size. When such a loss happens, so that the number of longitudinal veins becomes 6, 5, 4, or even less, the likelihood of regaining any is remote. On the other hand, in several orders the opposite kind of change has occurred and is equally obvious. The neuropteroid insects (lacewings, ant lions, and so on) have the number of veins increased by a multiplication of cross veins between the longitudinal veins. The Odonata (dragonflies and damselflies) have multiplied both the cross veins and the longitudinal veins to make a complex, netlike pattern. Yet through all these changes it remains possible to distinguish and to homologize the veins in even their most highly modified condition.

If an investigator is familiar with the morphology and adaptations of a group of animals, he can pick out many examples to show that the direction of evolution can be inferred from fairly complex mechanisms. In fishes we find some remarkable protractile jaws, as in the cichlid fish *Petenia splendida* (Fig. 38) from the Laguna de Petén, Guatemala. The mouth is larger than in related cichlids and can be opened widely for a sudden grab at prey. The premaxillary bones, united in the middle, bear a long median spine that lies freely in a groove all the way to the back of the skull. The enlarged and hooked maxillary provides leverage for thrusting out the premaxillary, and this slides forward, its direction controlled by the groove in which the spine rests. On comparing *Petenia* with its close relatives, we find that the genus *Cichlasoma,* with many species

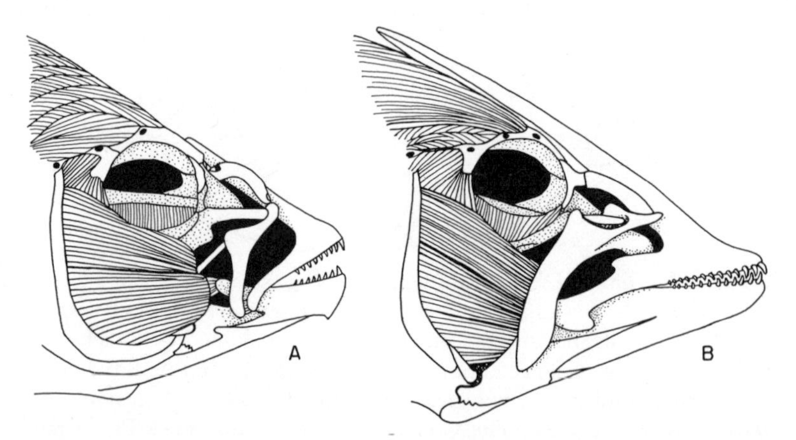

Figure 38. Jaw mechanism of two Central American cichlid fishes, showing structural stages in the evolution of an extreme type of protractile apparatus: (A) *Cichlasoma friedrichstahli;* (B) *Petenia splendida.*

in Central America, agrees with *Petenia* in structure but is less extreme. One species, *C. umbriferum,* has a premaxillary spine nearly but not quite so large as that of *Petenia. C. friedrichstahli* is similar but more like an average cichlid; the premaxillary spine reaches only as far back as the middle of the eye. It is possible, then, to set up a series of stages among living species that we may take to represent the "structural ancestry" of the unusual apparatus seen in *Petenia.* The expression "structural ancestry" does not mean that these stages *are* the actual ancestors of *Petenia,* but that they undoubtedly are much like, if not indistinguishable from, those ancestors. *Petenia* is in its own way unique among fishes and therefore represents an extreme of adaptation. On the other hand, the majority of Cichlidae have a protractile apparatus little different from that of many families of perchlike fishes. Therefore the latter condition is evidently generalized as far as cichlid jaws are concerned. Derivation of *Petenia* from among Central American species of *Cichlasoma* seems clear in view of their close similarity in other characters.

Just because a complex structure may carry within it evidence of its derivation, we can recognize cases of convergence or analogy that do not mean close relationship. The freshwater gars of the southern states, which are slender, predatory fishes with elongated upper and lower jaws and abundant teeth, are closely matched by a family of marine fishes, Belonidae, also called gars, or needlefishes. The resemblance in form is so striking that it would be easy to suppose that the needlefishes were derived from, or closely related to, the freshwater gars. But an examination of structures makes it clear that there is no relationship at all. The upper jaw in the freshwater *Lepisosteus* contains several bones, a marginal series corresponding to maxillaries, and premaxillaries, nasals, and frontals. The similarly shaped beak of a needlefish is composed entirely of the exaggerated premaxillaries. Bones of the dorsal surface of the skull, the shoulder girdle, the scales, the structure of the tail, and many other features are quite different. The freshwater gar is not even a teleost but belongs to the Holostei, a much older group of bony fishes.

Another example will be used to show the possibility of working out a succession of evolutionary changes in an ancestral line among animals in which the fossil record is lacking or inadequate. (It may be noted that several investigators are attempting to devise mathematical procedures for arriving at this kind of information.) The Centrarchidae are a family of familiar freshwater fishes limited to North America and perchlike in most respects. There are eleven genera of which most have one or two species each, but *Micropterus,* the black basses, contains five and *Lepomis,* the sunfishes, eleven or twelve. *Elassoma,* the pygmy sunfish, differs so sharply from the others that it is sometimes placed in a separate family.

In determining relationships there are certain principles (not "working hypotheses") that we may observe. First, many of the characters used in comparing members of a related group of animals can be arranged in pairs such that one member of each pair is more primitive than the other. Thus the presence of a lateral line on the body in sunfishes is certainly a primitive character of the family, having been brought as a heritage from among the large array of perchlike fishes; but the lateral line is absent on the body of *Elassoma.* Ctenoid scales (with a serrated margin) are also a primitive feature in sunfishes for the same reason as the lateral line, but we find in one genus, *Acantharchus* (mud sunfish), a contrasting character, cycloid scales. Some of the cheekbones have serrated edges in five genera (a widespread character of early spiny-rayed fishes), whereas in six genera these bones are normally smooth-edged. Several other pairs of characters may also be chosen in which it is possible to decide that one member of each pair is more primitive than the other.

The second principle is that within the group being studied all the members possessing the more primitive of a given pair of characters are directly related to each other with respect to this character (all received it from a common ancestor), but those with the less primitive character may or may not be directly related to each other, because it is possible for the less primitive (derived) condition to have appeared more than once independently. As an example of this, sunfishes in which some cheekbones are serrated must have inherited this from among the more primitive spiny-rayed fishes in which it occurs, but evidence from other characters shows unmistakably that the smooth cheekbones of *Acantharchus* were derived from the serrated condition independently of the smooth cheekbones of *Micropterus* and the other genera in which they occur.

Third, it is sometimes possible to distinguish three or more contrasting conditions and infer with full assurance which of these is the most, and the least, primitive. Thus the number of branchiostegal rays (bony supports in the opercular fold of fishes) in *Archoplites* and five other genera is primitive (there are seven), becoming six in the remaining genera except *Elassoma,* in which it is five. Comparison of the numbers of branchiostegal rays in many families and genera of teleosts indicates that the larger number is more primitive. But such conclusions, as already noted, cannot be reached with numerical characters by themselves but only when correlated with the evolution of nonnumerical features of the same animals.

It may now be clear that in the Centrarchidae there is a group of genera having certain relatively primitive characters but that in evolution of the family certain changes occurred before others did. For example, *Acantharchus* (as a genus) resulted from the loss of serrations on the cheekbones and of serrations on the scales while retaining seven branchiostegal rays. The number of branchiostegals dropped below seven in five genera, but in only one of these, *Elassoma,* was the lateral line lost from the body. There can be little question, then, in a given case, which of these events took place first, provided the distribution of characters is known throughout the family and that the primitive member of each set of characters is known. By such analysis in sunfishes it is possible to construct a diagram of the relationships and derivations of the genera and, in so doing, to show the sequence of changes in the evolution of the characters used. The *sequence* is real, but the absolute time involved cannot, of course, be known without an adequate fossil record.

Difficulties are often met in interpreting evolutionary relationships on physiological or behavioral grounds, but in many cases the difficulty vanishes when other kinds of evidence are applied. If, for example, a student who has no knowledge of the past history of reptiles, birds, and mammals is confronted with the question whether a sustained, high body temperature originated once, twice, or several times among these three classes, he is likely to start looking through recent physiological studies of temperature control, hoping to find clues on which he can base an opinion or seeking pertinent opinions of the authors, but he may assume that because the ancestors of these animals are now extinct, there cannot be any evidence from paleontology that would bear on the problem. Actually the evidence from paleontology is conclusive, and some of it is more than circumstantial.

Among today's reptiles the ability to control their temperature (or regulate the production and loss of heat) is variable and is generally slight. Much of it consists of habitat selection and choice of shelter or exposure to the sun, selection of hibernating sites, and so on. The orientation of a lizard's body to the direction of the sun is often used for either a warming or cooling effect, according to the time of day, ambient temperature, and so on. Metabolic rate and oxygen consumption depend upon body temperature, subject to hormones, especially that of the thyroid, which within fairly specific limits has a regulatory effect.

But modern reptiles have nothing to do with the ancestry of birds or of mammals, which lay far back among orders that are long extinct. That of birds was in the Thecodontia of the Triassic, the stem group of dinosaurs, pterosaurs, and Crocodilia; hence the nearest living reptilian relatives of birds are in the latter order, but even this connection is remote. Mammalian ancestry was in Triassic therapsids, and these in turn came, without question, from pelycosaurs, an order that goes back to the very oldest of all reptiles, in the early Pennsylvanian. For these reasons it is clear that endothermy, the maintenance of high internal temperature in birds and mammals, must certainly have originated separately in the direct ancestries of these two classes, that is, in the thecodonts on one hand and the therapsids on the other. It is associated with a way of life that is quite unlike, and demands far more sustained energy than, that of any reptiles that survive today. In mammals it is correlated with the presence of a diaphragm, which, under reflex control, provides for a rapid and sustained respiratory rate. A diaphragm functions in the absence of ribs in the abdominal region, a characteristic of mammals and of a few advanced mammal-like reptiles. It seems quite probable that this character, along with that of separation of the nasal from buccal passages in the same animals, signaled the development of a diaphragm and therefore of the increased rate of respiration, oxygen consumption, and metabolism that are the basis for endothermy.

In birds the mechanism is entirely different, for there is no diaphragm, and the lungs and their respiratory passages work on a different principle, involving the passage of air rapidly through into air sacs that are not respiratory and hence a continuous replenishment of oxygen-rich air in the lungs instead of vast numbers of blind alveolar pockets. Some orders of reptiles, past and present, have air sacs that similarly receive air by way of the lungs, but we cannot yet say with any assurance that thecodonts, pterosaurs, or dinosaurs had developed endothermy.

Thus the student's answer might properly be that several quite different approaches have been made among reptiles to the control of their temperature, mostly not very effective, but in birds one, and in mammals another, highly successful adaptation independently achieved this.

Another physiological trait that raises a similar question is that of excretion of urea (but little uric acid) in amphibians and mammals, but the production of uric acid crystals, with little water, in the urine of modern reptiles and birds. Does this mean that the ancestors of mammals must have excreted uric acid primarily, just because they were reptiles, and then switched to urea? Not at all, for the ancestry of mammals did not come through anything like modern reptiles but followed a wholly separate and in some ways very primitive line back virtually to the origin of reptiles among early amphibians. There is no reason to expect pelycosaurs or mammal-like reptiles to adopt uric acid excretion merely to demonstrate that they were reptiles; rather, the origin of this characteristic must have come among the early members of the other lines that gave rise to both the *modern* reptiles and the birds.

REFERENCES

Comstock, J. H. 1950. *An Introduction to Entomology*, 9th ed. revised. Comstock, Ithaca, N.Y.

Edmunds, G. F., and J. R. Traver. 1954. The flight mechanics and evolution of the wings of Ephemeroptera, with notes on the archetype insect wing. *J. Wash. Acad. Sci. 44*(12): 390–400.

Romer, A. S. 1949. Time series and trends in animal evolution. In G. L. Jepsen, G. G. Simpson, and E. Mayr (eds.), *Genetics, Paleontology and Evolution*. Princeton Univ. Press, Princeton, N.J.

———. 1966. *Vertebrate Paleontology*, 3rd ed. Univ. Chicago Press, Chicago.

15 · VERTEBRATE EVOLUTION

Thanks to a vast amount of work by paleontologists, especially in recent years, the past history of vertebrate animals is becoming quite clear in its main outlines. There remain, of course, innumerable problems and gaps in the record, some of which will probably never be filled. For further information than this brief chapter offers, Romer (1966) gives the most satisfactory account so far written.

More than 400 million years ago, in the Ordovician (see Fig. 32), certain sluggish, tadpole-shaped fishes, without jaws but with an armor of bony plates and scales, left their fragmentary remains in the sand of ancient seashores in Colorado. These ostracoderms are the oldest known vertebrates. They were followed by a diversity of ostracoderms during the Silurian and Devonian (Figs. 39 and 40), worldwide in distribution, evidently living close to or on the bottom, feeding on organic matter in the mud. Whether the larger number were in fresh or salt water is still disputed. They were probably not capable of fast or long-sustained swimming. There was no vertebral column but a cartilaginous notochord. Of paired fins there were none, or a pectoral pair in a few. The gills were in pouches between successive bony arches that were not jointed, and there were no movable jaws. The skeleton in almost all was bony, at least in the head. The internal cranial structure of several kinds of these primitive fishes has been preserved; it was shown by Stensiö and other Scandinavian workers that the brain, cranial nerves, and sense organs were closely comparable with those of the modern cyclostomes (lampreys and hagfishes).

The latter are two orders of limbless, eel-shaped parasitic fishes in which the skeleton is of soft cartilage, the gills are enclosed in pouches, and the gill arches are not jointed. But there are no scales and no bones. The notochord is persistent, as in ostracoderms. There is no doubt that both of these orders are descended from the extinct ostracoderms of the Paleozoic and probably along two different lines, although there is not yet complete agreement on the latter. As their soft skeletons are not readily

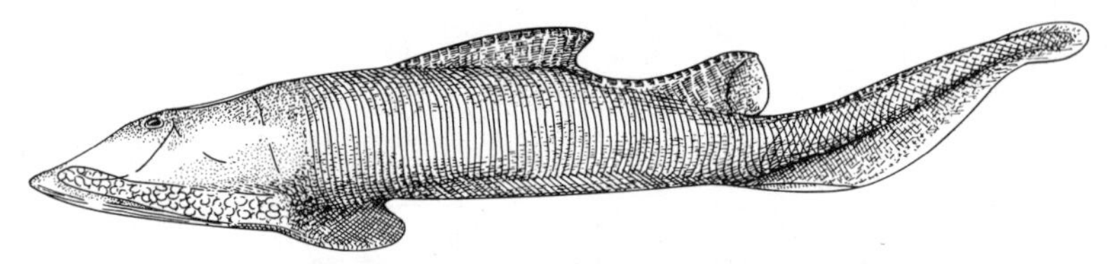

Figure 39. *Aceraspis robustus,* a Devonian cephalaspid ostracoderm.

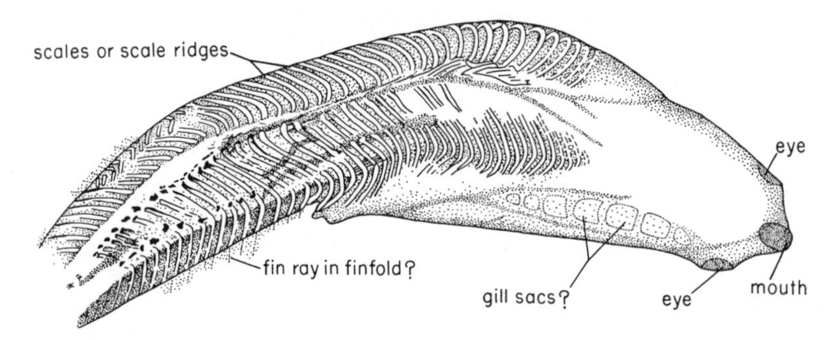

Figure 40. Supposedly unarmored (or perhaps larval) ostracoderm *Jamoytius kerwoodi.*

preserved, there is almost no fossil record, except for a recent discovery of lampreys in the Pennsylvanian. The anatomy of these two orders resembles that of different ostracoderms more nearly, in some ways, than the lampreys and hagfishes resemble each other. It therefore seems reasonable to include ostracoderms with them in a class Cyclostomata.

The earliest fishes that had jaws and teeth and in which the gills were on jointed arches with open slits between came in the Silurian. Although well armored still, these were more active creatures that pursued and ate other animals. The little, spiny-finned acanthodians were, from the beginning, quite unlike the ponderous placoderms (arthrodires and antiarchs; see Figs. 41 to 43). In both it is thought that the skeletal supports of the gills were internal to the gills, as in modern fishes, and that the first gill arch had been transformed to make an upper and lower jaw. Immovable pectoral spines, and similar spines for some of the other fins, were present before the development of freely movable fins or fin rays. We now suppose that acanthodians bore some relationship to the modern bony fishes of "normal" form but probably not that of direct ancestry. Some arthrodires attained a gigantic size, as, for instance, *Dinichthys* of the late Devonian Cleveland shale. It reached about 30 feet in length. Other placoderms were smaller, and some of these, losing nearly all their armor of dermal bone, appear the probable ancestors of sharks.

It is now clear that the sharks and rays (Chondrichthyes), on one hand, and the chimaeras or ghostfishes (Holocephali), on the other, were separately derived from among the less bony arthrodires of the Devonian

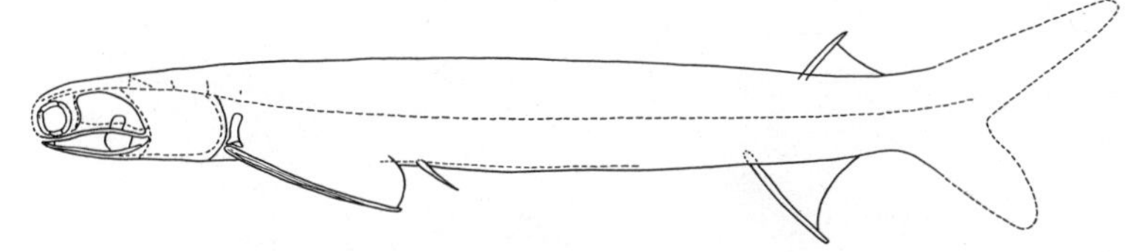

Figure 41. *Acanthodes,* a late survivor of the Acanthodii in the Permian.

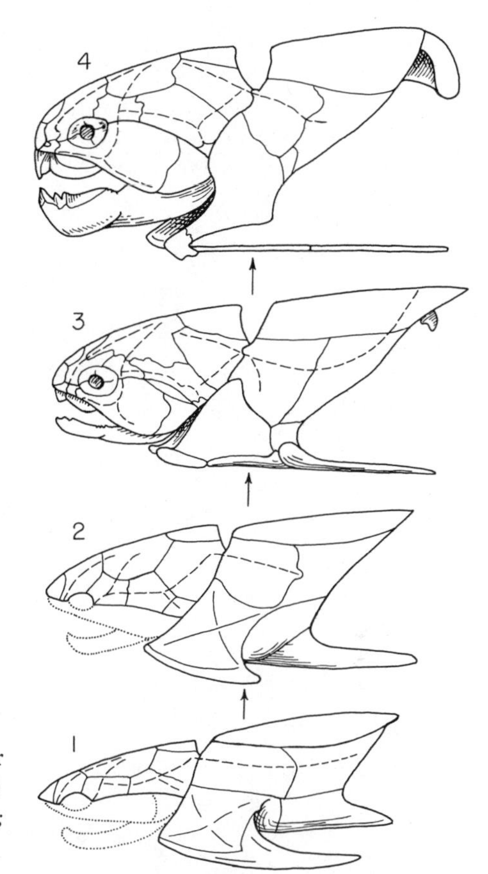

Figure 42. Evolution of the armor of arthrodires: (1) *Arctolepis;* (2) *Phlyctaenaspis;* (3) *Coccosteus;* (4) *Dinichthys.* (After Heintz.)

(Fig. 44). Until the last few decades it was taught that sharks, having a cartilaginous skeleton, are more primitive than the bony fishes, which have cartilage only in the larva. But it is now understood that the sharks themselves were descended from fishes that were bony, and their cartilaginous skeleton is simply the result of retaining throughout their lives a larval characteristic, failing to replace it by bone. Sharks, then, and rays, have nothing directly to do with the ancestry of any bony fishes. In fact, the latter came much earlier in the fossil record than did sharks, which put in their appearance no earlier than the middle Devonian. From this it

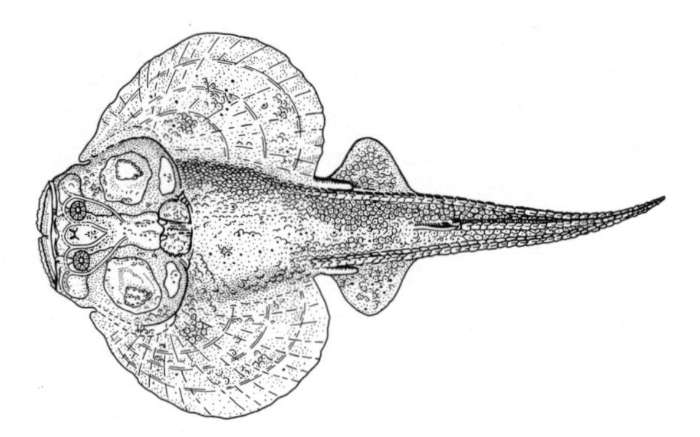

Figure 43. Flat, raylike arthrodire with reduced armor, the Lower Devonian *Gemuendina sturtzi.*

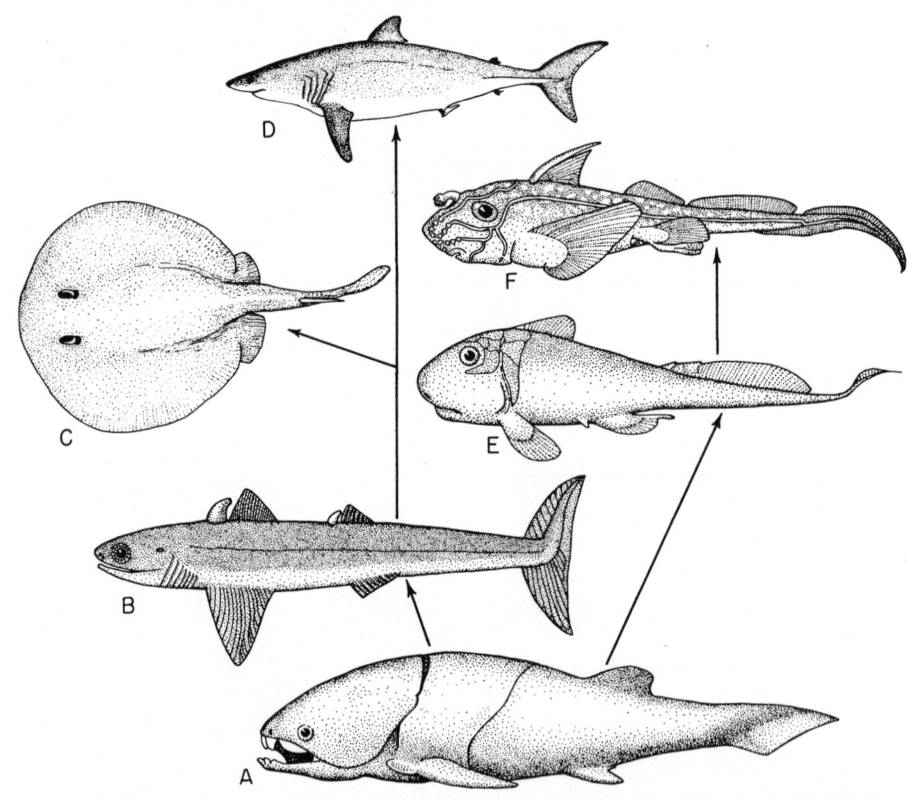

Figure 44. Oversimplified diagram of the relationships of the arthrodires and the sharklike fishes: (A) typical, armored arthrodire, *Coccosteus* (Devonian); (B) primitive shark, *Cladoselache* (Devonian); (C) Recent ray, *Urobatis;* (D) Recent shark, *Isurus;* (E) ptyctodont arthrodire, *Ctenurella* (Devonian); (F) Recent holocephalian, *Chimaera.* [Part (B) redrawn from A. S. Romer, *The Vertebrate Body,* 1962, courtesy of W. B. Saunders, Inc.]

can be seen that we should not expect the anatomy or physiology of sharks to shed any light on the ancestral condition of any higher group of vertebrates. As far as they are primitive in any respect, it is only by retention of features present among still older groups. Chondrichthyes did, however, make a place for themselves in the world and have remained with us to the present time as an active minority of modern fishes.

The class Osteichthyes contains all living bony fishes and began probably in the earliest Devonian (or Silurian) from something much like an acanthodian. Although the skull and shoulder girdle were encased in dermal bones the body and fins were scaly. The gills were covered by a hinged body operculum. Scale rows on the fins were gradually converted to flexible fin rays supporting a membrane. The skeleton was of bone in the later stages of development of individuals. The first known Osteichthyes from the Lower Devonian were already of two quite separate groups: ray-finned fishes (Actinopteri) and the lungfishes and lobefins (Sarcopterygii).

It is interesting that the notochord remained throughout life in early bony fishes and that the vertebrae at first barely enclosed it. Not until the middle of the Mesozoic did bony vertebrae become an actual replacement for the notochord in a mechanical sense; this probably is associated with the competitive development of greater speed and agility in swimming. Another point of interest concerns the fins. Those that we find now in modern fishes apparently originated more than once. The primitive arthrodires did not have movable paired fins, nor did the acanthodians. In arthrodires the pectorals were at first represented by huge, immovable bony spines projecting laterally, and the place of these spines was gradually taken (from behind) by flexible lateral folds in which muscles and skeletal supports developed. These fins became most sharklike in the later arthrodires, where bone had largely disappeared from the skeleton. It is from such unarmored, active, streamlined swimmers that the sharks of the mid-Devonian and later evolved.

In acanthodians, on the other hand, neither the paired nor the dorsal fins seem to have been at any time more than a series of sharp spines, each supporting a slight fold of skin. Yet the structure of the head and the scales of these fishes indicates that they were close to the source of Osteichthyes, and therefore the freely movable fins of the latter must have originated independently of those in the arthrodire-shark line. The stage of transition may some day be found, but it seems safe to postulate an active, lightly built, freshwater fish in which the spines were in the process of being replaced by a fold of scaly skin that was sufficiently free to undulate with the movements of the body.

The huge subclass called Actinopteri (ray-fins) includes the great majority of all known fishes. Even an outline of its history would be too much to present here, but, beginning with predatory stream dwellers, they radiated adaptively into all possible habitats of fresh water and the sea, undergoing certain progressive changes of structure and improvements in locomotion, culminating in the modern teleosts. A few of these are illustrated, showing modifications of form (Figs. 45 to 48). The primitive actinopterans, as shown by a few little-changed relicts, were lung breathers, having a pair of lungs as well as gills. These were reduced to one (still functional for breathing in two or three kinds of fishes); in teleosts as a whole the lung has become an air bladder with simply a hydrostatic function (offsetting by its buoyancy the weight of bones and teeth, so that the

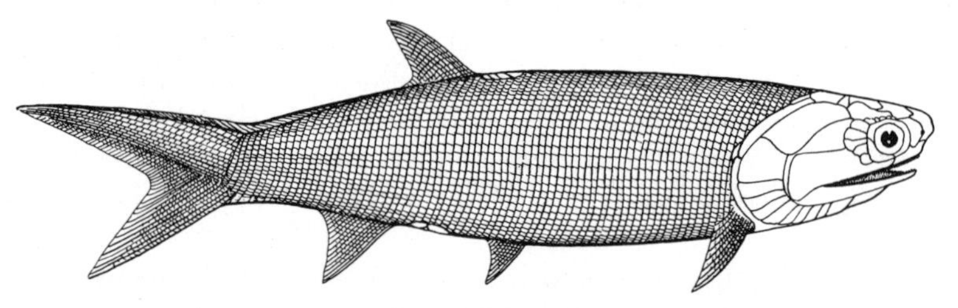

Figure 45. Permian ray-finned fish, *Palaeoniscus macropomus.*

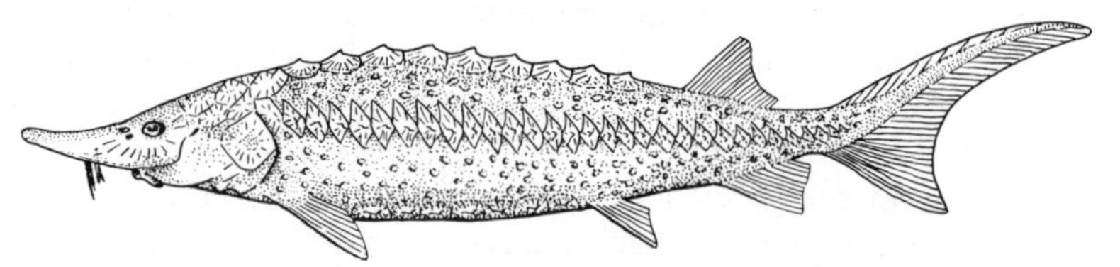

Figure 46. *Acipenser sturio,* a sturgeon (Chondrostei).

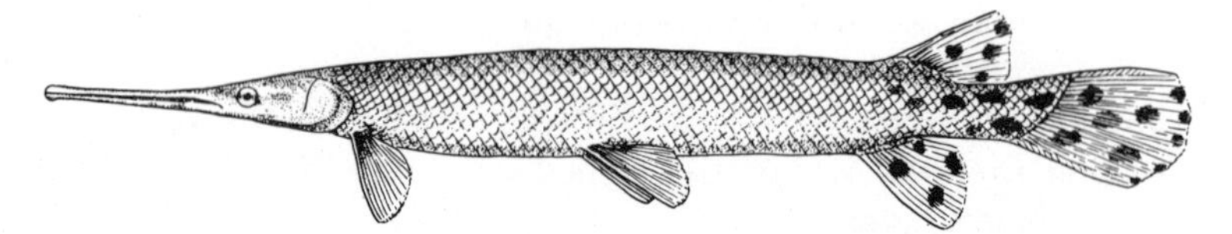

Figure 47. *Lepisosteus,* a freshwater gar (Holostei).

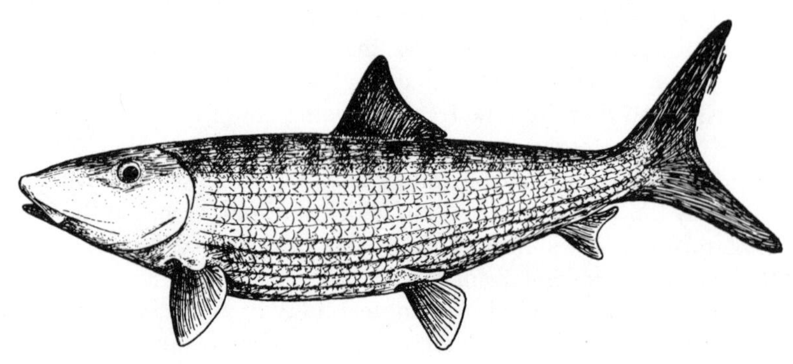

Figure 48. Ladyfish, *Albula vulpes,* a Recent marine teleost.

fish need make little or no effort to maintain its level in the water), and in some cases it serves as a producer or receptor of sound. The connection with the esophagus is lost in advanced teleosts, and in some families even the air bladder itself is absent. Thus the familiar idea that the lungs of land animals originated from the airbladder of a fish is not quite right; lungs were there first, in ancient bony fishes. This brings us to the other subclass of Osteichthyes, which likewise had (and kept) functional lungs.

Both lungfishes and lobefins are still with us today, the former represented by three genera in fresh water (Fig. 49), one in each of the southern continents. Living lobefins are a single genus, *Latimeria,* first found in the sea off the eastern coast of Africa. Its lung, incidentally, has degenerated to a storage organ for fat, so although it cannot be used for breathing, the hydrostatic function has not been lost, as fat is much lighter than water. But the older lobefins were in fresh water and breathed air, as indicated by the presence of nostrils. This group, Crossopterygii, is of great importance because of its relationship to Amphibia and higher animals (Fig. 50).

Formerly it was supposed, from the resemblance of lungfishes to salamanders (including a larval stage with external gills), that they must have been ancestors of Amphibia, but several specializations, such as fusion of their teeth into paired bony plates for grinding and union of the upper jaw solidly with the skull (not true of early amphibians), make this interpretation impossible, whereas several lobefins of the Devonian provide a pattern throughout their skeletal structure that agrees very closely with that of the earliest Amphibia.

The transition from lobefin to amphibian took place late in the Devonian. These four-legged animals appeared first in rocks of eastern

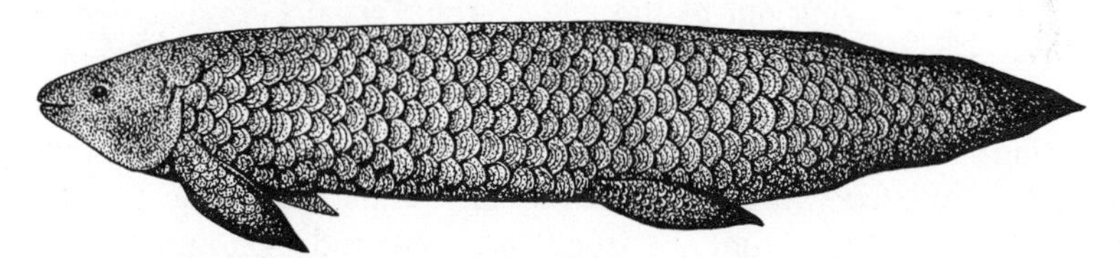

Figure 49. Australian lungfish, *Neoceratodus forsteri.*

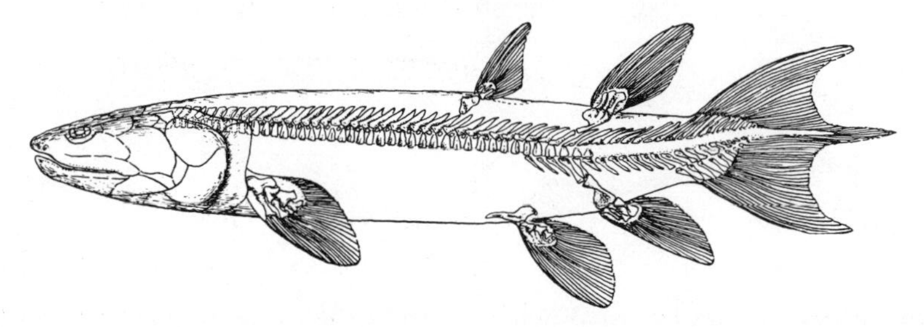

Figure 50. Late Devonian lobefin, *Eusthenopteron* (Crossopterygii).

Greenland (Fig. 51). They were fishlike in almost every respect, except that the limbs bore digits instead of fins and the limbs were jointed in the way usually found in amphibians. It is clear that these ichthyostegids lived in fresh water and could not have walked well on land, although they might have crawled a little way out of a stream. Instead of a good firm vertebral column there was an unconstricted notochord of cartilage. The vertebrae were of separate small bits of bone lying on the surface of the notochord (as in lobefins) and serving for attachment of segmentally arranged muscles but obviously not capable of providing mechanical support for the weight of the body. They were virtually identical with the vertebrae of Crossopterygian fishes. It is a point of great interest, however, that the earliest tetrapods should have appeared in the world at a time when ostracoderms were still present, when acanthodians and arthrodires were common, when sharks had not yet reached their maximum abundance, and when there were no teleost fishes, the group that has been dominant in the sea and fresh water ever since the middle of the Mesozoic. In fact, the general impression gained from a long-range view of vertebrate evolution is that the origin of each major division took place out of the more generalized members of the group that was ancestral to it. The only exception to this among the classes of vertebrates is the birds, which were derived from a progressive line of reptiles during the Jurassic.

The evolution of Amphibia is still not entirely unraveled, but a great many important facts about it are now clear. During the late Paleozoic there was a rapid radiation of many lines of Amphibia, primarily in fresh water but a few of them eventually becoming terrestrial. Most of the early amphibians, including the ichthyostegids, are known as labyrinthodonts. This large group reached its maximum in the Pennsylvanian and Permian, finally dying out at the end of the Triassic. Among the earlier members, the vertebrae consisted of separate small pieces of bone not completely replacing the notochord. The large bony skull frequently shows indications of grooves carrying the sensory canals of the lateral-line system, a heritage from fishes, and a clear indication of aquatic life. It now seems highly probable that the frogs and salamanders originated from a relatively primitive semiaquatic group of labyrinthodonts called Rhachitomi. This is on the basis of the structure of the skull and vertebrae. In

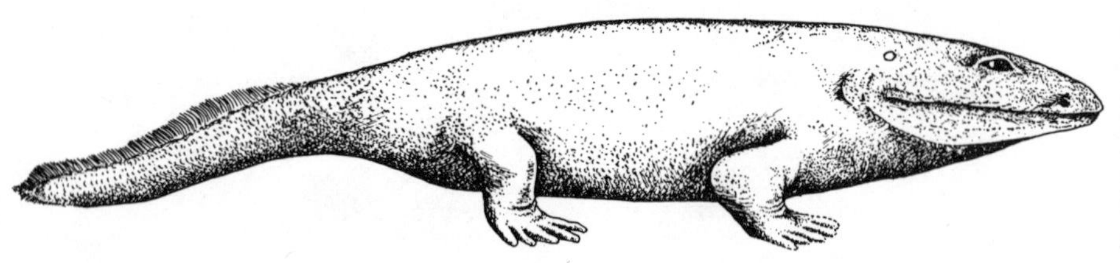

Figure 51. Earliest known Amphibian, *Ichthyostega*, of the late Devonian of East Greenland.

these two modern amphibian orders it may be that the adoption of a terrestrial habit took place independently, and certainly it had nothing to do with the development of land life in reptiles. The class Reptilia originated much earlier than either frogs or salamanders, out of one or perhaps two branches of the labyrinthodont order Anthracosauria (Fig. 52). These animals, incidentally, were entirely aquatic throughout their lives as far as we know, and it may well be that some of the earliest reptiles were also aquatic, although they had already achieved the skeletal structure of reptiles.

It is, and will probably always be, impossible to know whether the earliest reptiles laid their eggs on land and whether these eggs had a shell and a large yolk and the embryo had an amnion. As questions of embryology and physiology can rarely be answered by the fossil record, it is necessary to base most of our inferences about relationships upon the structure of the skeleton. In the case of reptiles, this means that the separate pieces composing the vertebrae of early Amphibia became united into single units in a different manner from that which they took in the case of the later labyrinthodonts and the modern Amphibia. Work published since 1955 has carried the history of reptiles back into the early Pennsylvanian with a strong presumption that the actual origin took place even before then. By the middle Pennsylvanian there had already diverged a line (the order Pelycosauria) that was to lead in the direction of mammals. Thus reptiles having a temporal opening on each side of the skull behind the orbit (as in mammals) were almost, if not actually, the earliest known members of their class. The aquatic anthracosaurs, before their extinction in the Permian, gave rise to at least two and perhaps more orders that have been variously classified as Amphibia or as reptiles. The source of turtles is still not definitely known. They appeared first in the Triassic. Probably all other reptiles can be traced to a common origin close to the beginning of the pelycosaurs but at first without temporal openings.

The Reptilia were the first class of vertebrates to occupy the land. After having solved the original problems of reproduction, early development out of water, and obtaining an adequate supply of food, probably in most

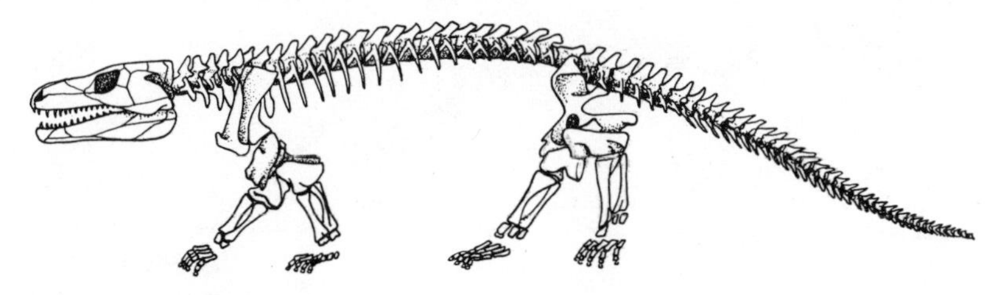

Figure 52. *Seymouria,* reconstruction of skeleton. (After White). It is a Permian amphibian close to the ancestry of reptiles.

cases insects, it seems that no obstacle remained to their conquest of the whole terrestrial environment. Habits and form that were superficially lizard-like appeared independently in several different groups and served as a general starting point for many excursions of adaptive evolution into quite different ways of life (Fig. 53). While the pelycosaurs and mammal-like reptiles were evolving during the Permian and Triassic, another major branch, the reptiles with two pairs of temporal openings, also radiated and gave rise to two subclasses, the lepidosaurs (for example, lizards) and archosaurs (crocodiles and their allies).

The first of these shows considerably diversity. It survives today as the lizards and snakes as well as the curious *Sphenodon* of New Zealand, a relict of another order, Rhynchocephalia. Among Lepidosauria, adaptive radiation produced limbless, burrowing forms several times, tree climbers frequently, and at least two or three gliders that took to the air in the manner of a flying squirrel. Lizards on more than one occasion became aquatic, as seen notably in the giant marine mosasaurs of the late Cretaceous (Fig. 54).

Even more successful in terms of diverse habit and structure was the subclass Archosauria, the "ruling reptiles." In certain primitive and somewhat lizard-like animals of the early Triassic, the thecodonts, there developed an extraordinary versatility based in part on agility and on the

Figure 53. *Kronosaurus,* a gigantic short-necked plesiosaur. Restoration based on a skeleton 42 feet long in the Museum of Comparative Zoology, Harvard University. Early Cretaceous of Australia.

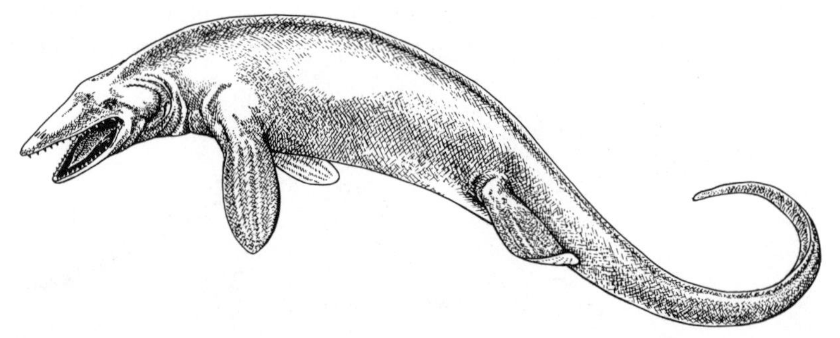

Figure 54. *Platecarpus,* a late Cretaceous mosasaur from the Niobrara Chalk of western Kansas.

habit of running on their hind legs. Within the thecodonts before the end of the Triassic there appeared large crocodile-like animals, the phytosaurs, and several running and arboreal types. This remarkable order served as the ancestral springboard for the evolution of two great orders of dinosaurs (Figs. 55 to 58), of the Crocodilia, which are the only Archosauria living today, of the Pterosauria or flying reptiles of the Mesozoic (Fig. 59), and finally of the birds. It should be emphasized that the birds and pterosaurs were independent of each other and both originated, separately, from among the arboreal thecodonts.

There is no close resemblance between pterosaurs and birds except that they flew. The wings and general structure of the body were distinctly

Figure 55. *Ceratosaurus,* a large carnivorous theropod dinosaur, late Jurassic.

Figure 56. *Brachiosaurus,* largest of the sauropod dinosaurs, late Jurassic.

different. They were contemporary with each other for many millions of years, from about the middle Jurassic, when the first known bird appeared, to the end of the Cretaceous, when the pterosaurs finally became extinct. It is possible that competition by birds aided in bringing about the extinction of the flying reptiles, but this is not necessarily the reason, nor does it account for the disappearance at about the same time of the last dinosaurs, mosasaurs, plesiosaurs, ichthyosaurs, and other important groups of Mesozoic reptiles. The cause for this great (but probably not rapid) extinction remains unknown.

Figure 57. *Corythosaurus,* a crested dinosaur of the late Cretaceous.

Figure 58. *Monoclonius,* a late Cretaceous ceratopsian dinosaur. (Redrawn from E. H. Colbert, *Evolution of the Vertebrates,* 1955, courtesy of John Wiley & Sons, Inc.)

Figure 59. *Pteranodon,* a great flying reptile of the late Cretaceous Niobrara Chalk, western Kansas.

The fossil history of birds in the Jurassic and Cretaceous (Fig. 60) is scanty but sufficient to show a rapid evolution from the long-tailed reptilian type, *Archaeopteryx*, to more modern types that had reduced the tail to a short stump but still possessed teeth, and finally the appearance before the end of the Cretaceous of several of the recent orders of birds. Then, in the Cenozoic, all the modern familiar orders seem to have been present, along with various seemingly experimental types that have come and gone. These include gigantic flightless birds of several different orders, such as the moas of New Zealand and the elephant birds of Madagascar, as well as at least one immense flying sea bird, *Osteodontornis*, much larger than any others that are known.

The birds as a class show certain parallels to, but of course no direct relationship with, the mammals. The most important of these is the maintenance of a high metabolic rate accompanying control of body temperature. This entails rapid oxygen consumption and is related to the extraordinary activity manifested throughout the lives of almost all birds. Of course, there is also a high development of the central nervous system, especially the cerebellum, the optic lobes, and the cerebral hemispheres, in correlation with the control of motor activity, the senses of sight and equilibrium, and the development of complex behavior.

Returning to the Permian, we find that in the meantime the pelycosaurs gave rise to both carnivorous and herbivorous families, some of them relatively large animals, and that the carnivores in turn by the early Permian were the source of a new order, Therapsida, the mammal-like reptiles.

Figure 60. *Hesperornis*, a large, toothed, diving bird of the late Cretaceous Niobrara Chalk, western Kansas.

Many hundreds of species of this remarkable order are known. They persisted in small numbers into the Jurassic but most lines had disappeared before that. To deal with the classification and adaptive radiation of these reptiles in any detail would require far more space than is available, but there was a major division into an herbivorous suborder, the anomodonts, and a carnivorous suborder, the theriodonts (Fig. 61). Both of these were of worldwide distribution and produced numerous bizarre types. The theriodonts, in several more or less parallel lines, became increasingly like mammals in regard to such characteristics as the teeth, the development of a secondary palate, two occipital condyles, an upright position of the limbs under the body rather than spread laterally, and other modifications of the skeleton. It is impossible to know when or whether the mammal-like reptiles developed hair instead of scales, whether any of them were viviparous, and whether they fed their offspring with milk, but all these are possible. It is virtually certain that at least the later members had some degree of temperature control, which implies a high metabolic rate and rapid oxygen consumption, and it now appears that the transition from reptiles to mammals took place in possibly two or three different lines of therapsid reptiles at about the end of the Triassic.

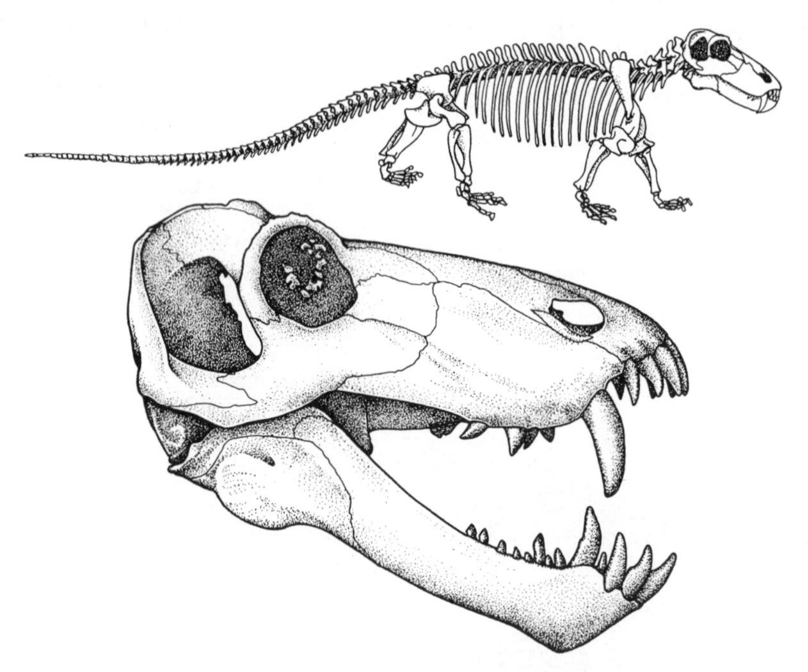

Figure 61. *Titanophoneus,* a primitive mammal-like reptile of the late Permian of Russia. Above, restoration of skeleton, actual length about 8½ feet; below, skull, actual length about 16 inches. (Redrawn from J. Orlov, *Traité de Paléontologie,* 1961, courtesy of Masson et Cie.)

We cannot speak definitely about physiology or reproduction, so it is necessary to rely upon a chosen skeletal characteristic or a combination of such that can be found in fossils, to mark the point of transition. In this case, the lower jaw of a mammal has only a single bone, the dentary, on each side and this is articulated with the skull in a glenoid fossa on the squamosal bone, whereas in reptiles there are several bones in the lower jaw, and the articulation is between the articular and quadrate. In mammals, the articular and quadrate have been converted to sound-transmitting bones of the middle ear, the malleus and incus, respectively.

Mammals of today comprise two very unequal groups, the egg-laying monotremes, and the viviparous Theria, which are all the rest. Monotremes, the duckbilled platypus and spiny anteater, are limited to Australia and New Guinea, have no fossil record, but retain certain primitive reptilian characters of the skeleton and reproductive system while being highly specialized in regard to skull, jaws, and teeth (no teeth present in the spiny anteater and only vestiges in the duckbill). Evidently, then, they stand at the end of a very divergent line that perhaps originated separately among the therapsid reptiles. Concerning several little-known early divisions of the remaining mammals we need say nothing, but notice the subclass Theria, which separated during the Cretaceous into marsupials (Metatheria) and placentals (Eutheria).

The most satisfactory evidence concerning this separation has been found recently at several localities in western North America (Clemens, 1968), where early opossums and primitive insectivores are represented by jaws and dentitions that are closely similar. At a time when dinosaurs were declining but still present, the placentals were beginning their differentiation into stocks that would give rise, a little later, to the primitive carnivores, primates, and ungulates, but in the late Cretaceous these were still not sufficiently distinct to be removed from the order Insectivora.

The story of the spread of ancestral marsupials into South America and presumably around the shores of the Pacific to Australia, and of their adaptive radiation in these two island continents during some 60 million years of the Cenozoic, cannot yet be told in full. They apparently have been absent from the rest of the world since the Paleocene at least, while they produced among themselves various carnivores (Fig. 62), insect eaters, and herbivorous types that were completely different in those two areas. Although the South American marsupial fauna competed with several ancient placental groups that also evolved there, they were not prepared for the encounter with advanced mammals (Fig. 63) that entered in the Pliocene when the connection with Central America was reestablished, and all South American marsupials except a few species of opossums became extinct.

Meanwhile, following the extinction of dinosaurs and many other Mesozoic reptiles at the end of the Cretaceous, placental mammals radiated extensively, spreading over most of the world, in environments that differed little from those of today except for a generally milder, more uniform climate. The other groups of vertebrates, the invertebrates, and vegetation were essentially modern. Intelligence, insofar as it is shown by increasing size of the cranium and brain, was not characteristic of any particular advancing line of placental mammals but evolved in several

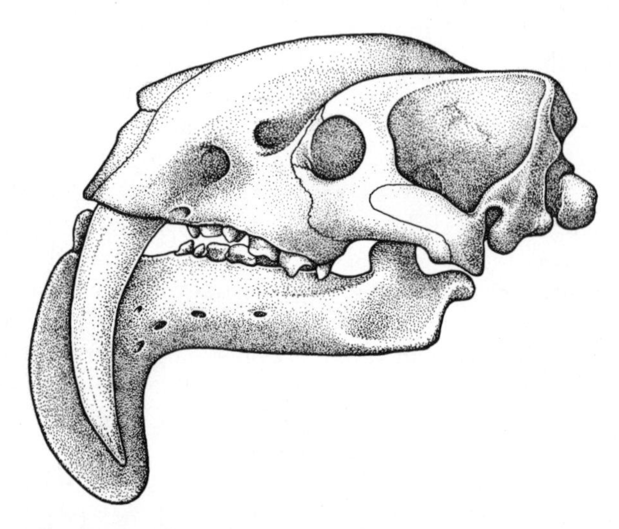

Figure 62. Skull of the South American Pliocene saber-toothed marsupial *Thylacosmilus atrox*, a remarkable case of convergence with the saber-toothed cat, *Smilodon*.

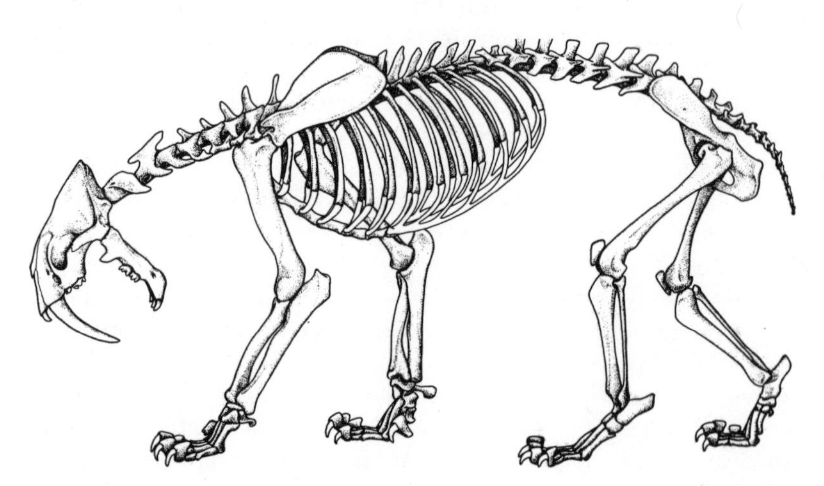

Figure 63. Saber-toothed cat of the Rancho La Brea tar pits, *Smilodon californicus*.

groups independently. Ancestral primates had no better brains than ancestral rodents, horses, elephants, whales, carnivores, or any other line.

A great deal is known about placental evolution in the Cenozoic, more than for any other group of comparable diversity, but new discoveries continue at an increasing rate even while the better known branches of the tree are being traced to their common sources.

REFERENCES

Clemens, W. A. 1968. Origin and early evolution of marsupials. *Evolution* 22(1): 1–18.

Colbert, E. H. 1955. *Evolution of the Vertebrates.* Wiley, New York.

Romer, A. S. 1966. *Vertebrate Paleontology,* 3rd ed. Univ. Chicago Press, Chicago.

16 · EXAMPLES OF EVOLUTIONARY LINES

Although much of the current interest in evolution among biologists is concerned with the processes by which the genetic characters of populations change, more and better work than ever before is being published on the long-range results of evolution, that is, on the phylogenetic relationships of ancestors and descendants in various groups of organisms. The nature of this work and its conclusions are often somewhat taken for granted by those who do it and tend to be disregarded by others who are not directly concerned. Therefore it seems advisable to discuss in this chapter some examples of the kind of information that is available and the way in which it can be used to obtain a broad picture of evolutionary relationships.

The two examples used require different approaches because the kind of evidence available is different and the reasoning cannot have the same basis. The first example chosen is a widespread and abundant order of freshwater fishes, concerning which the fossil record is far too scanty to give any help; nor does embryology provide useful information, mainly because there has not been enough comparative embryological study of the fishes concerned. We must rely almost exclusively on morphological comparisons supplemented by ecological and geographical considerations. Yet the result is quite clear as far as the relationships among major divisions of this order are concerned. The second case, that of the horse family, differs from the preceding in that nearly all the evidence, and it is remarkably complete, comes from paleontology, although, of course, this is mainly morphology relative to time. In the present chapter, these two examples of phylogenetic study are not complete in regard to the evidence or final as to the conclusions, but they illustrate the kind of information that is available and the ways in which it can be used.

THE FISH ORDER OSTARIOPHYSI
(Catfishes, Carps, Minnows, and Their Relatives)

The Ostariophysi constitute a natural group containing the great majority of freshwater fishes of the world, about 40 families and 5000 species, as well as two families of secondarily marine catfishes. There are fossils of some members of the order, but they are scarce and fragmentary and do not add to our knowledge of relationships. In all these fishes there is a series of specialized bones, the Weberian ossicles, derived from ribs and parts of the first four vertebrae, connecting the air bladder with the inner ear (Fig. 64). No other fishes have this peculiar adaptation, which therefore marks the order as a specialized branch among the teleosts and not ancestral to anything else. Although the Ostariophysi are greatly diversified they share several other characteristics that are in general those of the more primitive teleost fishes. The air bladder usually has a duct to

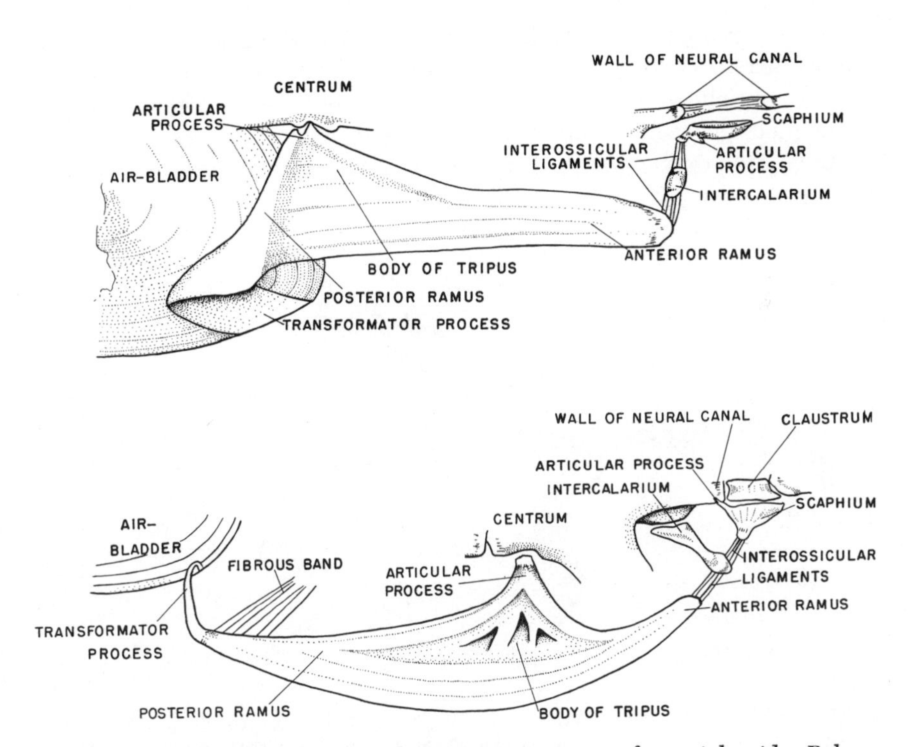

Figure 64. *Weberian apparatus, diagrammatic,* as seen from right side. Below, as in carplike fishes (Cyprinoidea); above, as in catfishes (Siluroidea). (After Krumholz.)

the esophagus. The pelvic fins, if present, are in the abdominal position. The fins are usually without spiny rays, or the dorsal may have one or two and the anal and pectorals one each. There is commonly but not always an adipose fin, that is, a small fleshy fin, in the position of a second dorsal but without rays.

The Ostariophysi can be divided conveniently into two suborders, one of which, the Siluroidea, contains the catfishes, with about 28 families and many hundreds of species. The other suborder, Cyprinoidea, is even more numerous and diverse. It includes the tropical characins of South America and Africa, the gymnotids, which are South American eel-like fishes, and the very large group of minnows, carps, and their relatives. (Fig. 65).

Among the Cyprinoidea, the body is usually scaly, sometimes naked, but never with bony plates. Certain skull bones characteristic of most fishes are present—the parietals, symplectic, and subopercular—and there are several other generalized characters that are common to most primitive teleosts. It is convenient to classify the members of this suborder in three superfamilies. One, known as the characins, contains about six families and nearly 1000 species in tropical and subtropical America and Africa. In comparison with the other two superfamilies, this one has a more generalized combination of characters. The dorsal, pelvic, and anal fins are present. The anal opening is in the usual posterior position. The lower pharyngeal bones are normal and the upper jaw is almost never protractile. Both the premaxillary and maxillary bones serve as the fixed margin of the upper jaw and carry teeth.

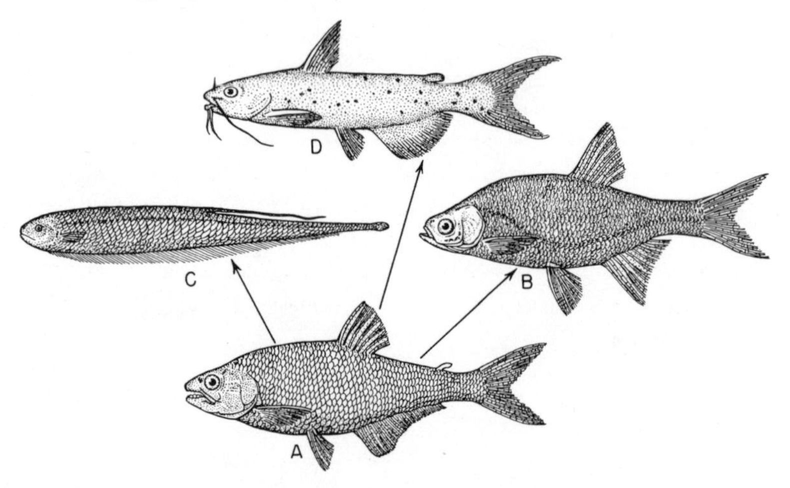

Figure 65. Relationships among the major groups of ostariophysan fishes: (A), characin, *Brycon;* (B), cyprinid, *Notemigonus;* (C), gymnotid eel, *Protergus;* (D), catfish, *Ictalurus.*

On the other hand, the gymnotid "eels," with four families containing about 35 species in South and Central America, differ greatly in having no dorsal fin or at best a threadlike adipose fin, no pelvics, the caudal absent or vestigial, and the anal fin very long, often originating in front of the pectorals. The anus is usually under the head and never behind the pectorals. The gymnotids agree with characins, however, in having a non-protractile upper jaw and normal lower pharyngeal bones. One member of this superfamily is the electric eel. Evidently the gymnotids are a highly specialized and peculiar division of the Ostariophysi which could hardly have been derived from any other group than the characins and have certainly not led to any other.

The largest of the three superfamilies contains four families of minnows and carplike fishes. In these, the lower pharyngeal bones are enlarged and hook-shaped and the upper jaw is freely protractile, having the margin of the mouth formed only by the premaxillary, which can be moved forward and back by means of a lever system composed largely of the movable maxillary behind it. But these fishes are usually scaly, of normal shape, and have soft-rayed fins in the usual positions. The minnows are the largest family of fishes, containing some 2000 species in Europe, Asia, Africa, and North America. They are most numerous and varied in south-eastern Asia. There are also suckers, mostly in North America, and two families of loaches in Europe and Asia, as far east as Java and Borneo.

To review the Cyprinoidea, evidently the characins contain the primitive stock, out of which independently arose the gymnotid eels and the minnows, the latter being much more successful in terms of distribution and adaptation to varied circumstances.

There are almost 2000 species of catfishes, belonging to the suborder Siluroidea and grouped in about 28 families. The body is naked or covered with bony plates instead of scales. The symplectic, subopercular, and parietal bones are absent. The second, third, fourth and sometimes fifth vertebrae are fused. The maxillary bone is nearly always rudimentary, supporting a barbel, or "whisker." The upper jaw is not protractile. It seems therefore that there is no other group from which catfishes could be derived except the primitive stock of the order, the characins. In one small family of catfishes living in the streams of Chile and western Argentina, the maxillary bones are well developed and have teeth. This is then the primitive family of catfishes, for in all others the maxillary is vestigial and toothless. Catfishes occur in the greatest number and variety in the fresh waters of South America, but a few kinds range up into Central America, and there is one family containing about 36 species in North America east of the Rockies. Africa and the southern and eastern part of Asia are well supplied. Two families of catfishes are largely marine, living in warm seas and able to enter brackish or fresh water; one family in southern New Guinea has evidently been evolved from a marine group and lives exclusively in fresh and brackish water.

It is clearly feasible, with sufficiently complete information about morphology and distribution, to analyze a family or an order of animals and determine in several ways what is the primitive stock or the primitive combination of characters that must have occurred in the common ancestry of the group. It is also possible, if several distinctive characters are available, to recognize the derivatives of the primitive stock and to understand whether they are derived separately from it or from one another. All of this is entirely practical and often relatively simple even when no fossil record is available whatever, although it is a favorite saying among some biologists that any phylogeny in the absence of a fossil record is pure speculation and that it is highly controversial even when there is a fossil record.

THE HORSE FAMILY

The horses of today and their relatives, the zebras and wild asses, belong to the genus *Equus*, which stands at the end of a long line of evolutionary history extending back about 50 million years to the beginning of the family Equidae in the early Eocene. This history is probably unique in its completeness, for it is possible to trace back the ancestry of modern horses to preceding genera and species, and these to their ancestors, and so on with scarcely any gap in the record. The evolution of the horse family is not in a single line, but, as with other histories, it shows many branches that diverge from one another. All have led to extinction except that represented by the modern genus *Equus*. The most satisfactory general source is Simpson (1951).

The completeness of the record introduces an unusual factor in its interpretation, for in going from the recent horses to those of a slightly but not much earlier time, there is no need to search among mammals of various types for one that could be considered ancestral to a horse. Instead, the earlier horses themselves are clearly, unmistakably recognizable among all other mammals contemporary with them, simply because they do not differ greatly from horses of today. Likewise, in comparing fossils of a particular age to those of either a later or an earlier time not far removed, the same problem is solved in the same way by recognition of animals that are very little different from those already being considered. The same principle holds for the comparison of any particular parts of these animals, for example, the pattern of the molar teeth, the shape of the skull, or the arrangement of bones in the feet. Through a succession of ages in which horses are known, the result is a sequence of changing characteristics in which the stages differ only slightly from one another. In fact, through much of this history the changes are of scarcely more than a statistical nature, such as would be seen in overlapping populations.

Long before the fossil record had become known as completely as this, it seemed that the changes in characteristics of the horse family beginning in the Eocene represented a very regular progression in several features, especially the size of the animal; the proportions of the limbs, body, and head; the number of digits on the feet; the form and pattern of molar teeth; and the size and proportions of the brain. The closely coordinated regularity of these changes suggested that the horses were an excellent example of evolution in a straight line. There are two aspects to this concept of *orthogenesis.* One is that the changes seemed to proceed without deviation and might lead in the course of time to an extreme that was no longer able to survive because it had passed the limits of adaptation. The other is that this straight line of evolution could be explained by an innate controlling force in the genetic makeup of the animals that produces gradual change regardless of circumstances or adaptive advantage, even though it lead to extinction. But fuller understanding of evolutionary processes and the more complete elucidation of the fossil record of horses have shown that neither of these aspects of orthogenesis is true or is necessary from the evidence, and the whole orthogenetic principle is a fallacy based on insufficient knowledge.

The oldest genus of horses, once widely known as *Eohippus,* but by priority of name correctly *Hyracotherium* (Fig. 66), appeared in the early Eocene of both Europe and North America at a time when the two continents were connected by land across what is now Bering Strait. Several species of these small animals are known, varying in height from less than 1 foot to about 20 inches. The front feet had four toes, each tipped with a small padlike hoof, the thumb being absent and the longest toe the second (which would be third if the thumb were present). The hind foot had three toes with hoofs, the first and fifth being reduced to tiny vestiges. The body, instead of straight, was somewhat arched and the actual tail (in contrast to the hairs that it may have carried) was relatively longer and heavier than in any later horses. On the head, in contrast to modern horses, the muzzle in front of the eyes was no longer than the cranium behind the eyes. The brain was small and without convolutions but quite similar in general appearance to that of the modern opossum. The teeth of *Hyracotherium* numbered 44, as in the early insectivores and primitive members of several other orders of mammals. The four premolars were recognizably different from and simpler in pattern than the three molars behind them. The animal evidently fed on soft leafy vegetation and it differed very little from some members of the extinct order Condylarthra, from which several orders of herbivorous mammals evolved.

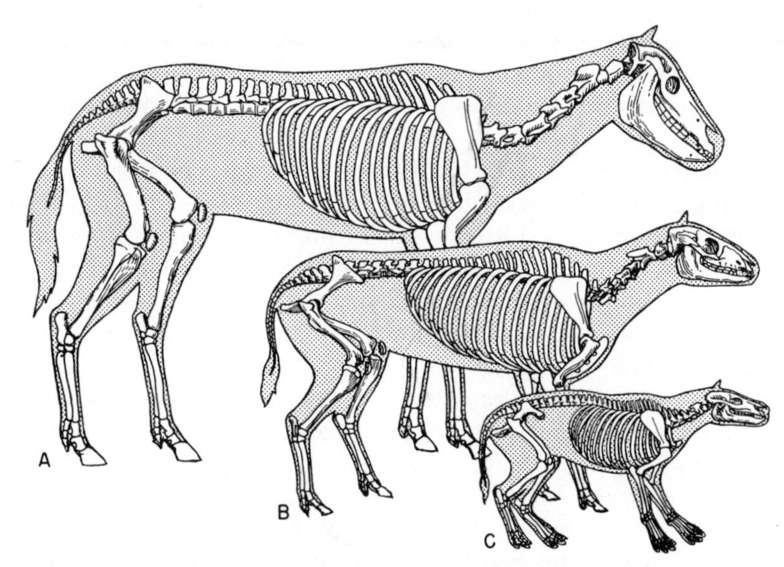

Figure 66. Early Tertiary horses: (A) *Merychippus,* Miocene; (B) *Mesohippus,* Oligocene; (C) *Hyracotherium* (= Eohippus), Eocene.

There is a small but definite gap in our knowledge between the earliest horses and the earlier Condylarthra of the Paleocene, immediately preceding the Eocene. During the middle and late Eocene the connection between Asia and North America was again submerged, with the result that horses of the Old World and New World went their own ways in evolution. Those of Europe died out finally at the end of the Eocene, but in North America the line was more successful. The genera *Orohippus* and *Epihippus* succeeded *Hyracotherium* in the middle and late Eocene, respectively, with some changes in the characteristics of premolar and molar teeth but little progress in size or other features (Fig. 67).

Following *Epihippus* in the early and middle Oligocene there came a somewhat larger horse, *Mesohippus,* having longer legs, three toes on each foot, the bones of the toes and the hoofs being larger and stronger than in the Eocene horses, the body straighter, and the tail relatively shorter. The height was about 2 to 3 feet. The brain was distinctly larger and the cerebral hemispheres somewhat convoluted. The part of the skull anterior to the eyes was slightly longer than the cranium behind the eyes. All the cheek teeth except the first premolar now had essentially the same pattern of cusps and ridges but were still adapted to browsing on soft vegetation rather than harsh grass, which was not yet available.

With relatively little structural change but some increase in size, the characters of *Mesohippus* blended into those of the late Oligocene and early Miocene genus *Miohippus,* although this involved merely a small detail in the structure of the molar teeth and a slightly more rigid contact

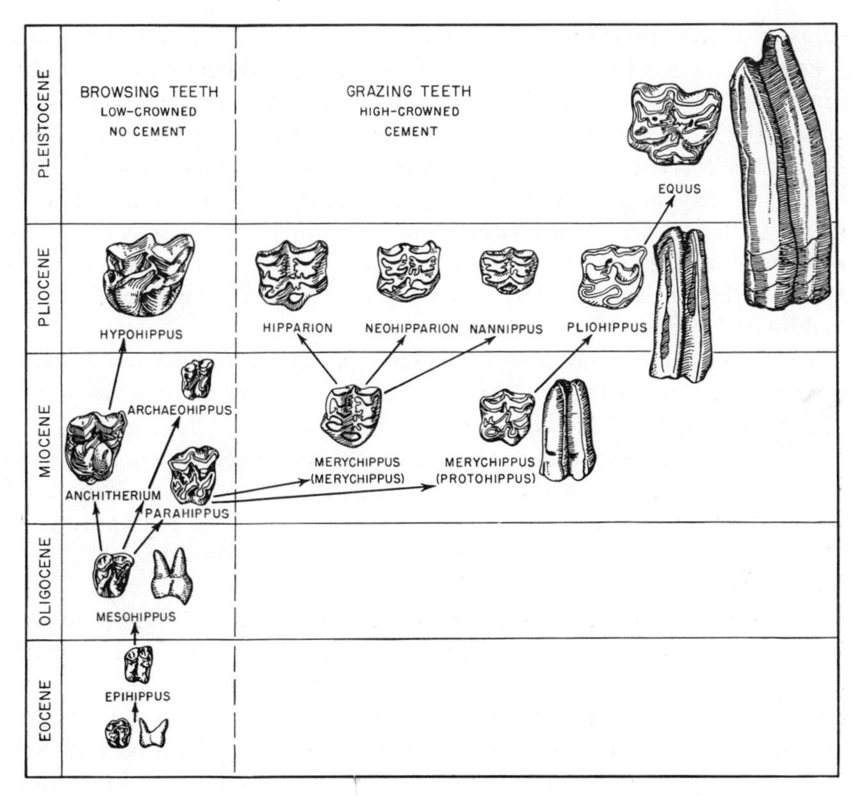

Figure 67. Evolution of right upper molar teeth of horses. (Redrawn from Simpson.)

among the bones of the hind foot. There was then a divergence of evolutionary lines in different directions from *Miohippus.* At least two of these lines retained relatively simple, short-crowned, browsing-type teeth. One branch, *Archaeohippus,* was reduced in size to something like that of the earlier Oligocene horses and then became extinct. Another short-crowned group continued to increase in size through the Pliocene. It is probable that these animals continued to be forest dwellers, whereas the third line, starting with *Parahippus,* may have initiated the habit of grazing in open grassy fields, which apparently were beginning to appear in some parts of the world during the Miocene. *Parahippus,* although still three-toed, developed what are known as hypsodont molar and premolar teeth, capable of continuing growth at a rate compensating for the wearing down of their grinding surfaces and also resisting such wear by development of complicated ridges of enamel to do the grinding. The special significance of this is that grass is unusually abrasive, as it contains silica and, for that matter, dust and grit near the ground, and only animals that were adapted to chewing great quantities of grass could take full advantage of the opportunity provided by this successful family of plants.

Parahippus was followed in the middle and late Miocene by the highly successful grazing genus *Merychippus*. This animal resembled fairly closely the modern horses in its proportions but had not yet attained their size. In the forelegs, the radius and ulna became fused into a single bone instead of remaining separate as in more primitive horses. The middle toe of each foot was much larger than the two lateral toes and carried most or all of the weight. At least five branches of the three-toed horses appeared in the late Miocene and Pliocene but had disappeared by the middle Pleistocene.

Meanwhile, in the Pliocene the genus *Pliohippus* became one-toed by the complete loss of the two reduced side toes. From among the widely distributed species of *Pliohippus* there arose in the late Pliocene the genus *Equus* and a peculiar South American group of three genera, whose ancestral stock was able to reach that continent when the isthmus of Panama arose from the sea. Thus, for a while, single-toed horses lived on all the continents except Australia. The Pliocene invaders of South America were replaced there, perhaps competitively, by species of *Equus* during the Pleistocene. It was not until the very late Pleistocene that horses became entirely extinct in the western hemisphere, and they did not appear again until they were brought from Europe by the Spanish explorers. The wild species of *Equus* of historic times are in Africa, eastern Europe, and central Asia. Domestic horses of Europe and Asia probably originated from two or three sources somewhere in western Asia.

REFERENCE

Simpson, G. G. 1951. *Horses*. Oxford University Press, New York; also Doubleday (Anchor), New York, 1961.

17 · MAN AND THE PRIMATES

Primates are the monkeys, apes, man, and certain other, more primitive groups such as lemurs and tarsiers. Most of them can be distinguished from other mammals by the grasping form of the hand and foot, in which the thumb and big toe oppose the other digits, and by nails rather than claws on at least some of the digits. Associated with grasping hands, the eyes look forward, with a large overlap of their visual fields, allowing close binocular vision. These and some other features are adaptations for life that is primarily arboreal, but they are not all present in all primates.

Primates today are tropical animals living in well-watered, forested country. This includes man, who, however, has been able to spread far beyond the tropics and become worldwide. The living primates most similar to man are the apes, of which two kinds, the gorilla and chimpanzee, live in equatorial Africa; the smaller gibbons are Oriental, including the Malay region, and the orangutan is in Borneo and Sumatra. All are fairly well known to people who visit zoos regularly. Apes, of course, are tailless, the arms are longer than the legs, and there is a more or less human expression in the face, but the brain is smaller than that of man, reaching no more than about 600 cubic centimeters in a large gorilla.

Of monkeys, there are two major groups, quite different from each other. One is the Old World monkeys of Africa and Asia, extending eastward to, but not beyond, Borneo and Japan. In general, the arms and legs are of the same length. There is usually a long tail but it is not prehensile. The group includes baboons, which have a short tail held partially upright. A number of Old World monkeys are not limited to trees but are partly terrestrial, and this is especially true of baboons. The so-called "Barbary ape" is a monkey with no tail that lives on Gibraltar, where there are few trees to climb. In general, the Old World monkeys have a narrow septum between the nostrils and the number of teeth is 32, as in apes and man. The other distinct group of monkeys includes two families in South and Central America in which the septum between the nostrils is relatively wide. Several of these have tails that are prehensile, especially the spider monkeys, and all are highly adapted for arboreal life. It is possible that they have originated from earlier, nonmonkey primates independently of those in the Old World. The number of teeth, 36, is a more primitive feature. Marmosets, howlers, capuchins, and spider monkeys are included in this group.

In addition to monkeys, the order includes several other more primitive members in the Old World tropics. Most familiar of these are the lemurs, of which many kinds live in Madagascar and just a few in Africa and India. They are variously specialized but might be described superficially as being between monkeys and squirrels in appearance. The face is longer than that of monkeys and the eyes somewhat more lateral. There is usually a bushy tail. The hands and feet are equipped for grasping. The size ranges from that of a squirrel to one or two recently extinct kinds that were nearly as large as a man. The lemurs are all arboreal, usually nocturnal, and feed on fruits and insects.

In the forests of the far east, from Sumatra to Borneo, Celebes, and the Philippines, there lives the small, nocturnal tarsier, which has enormous eyes, disks on its fingers and toes, and is distantly related to lemurs. Also in southeastern Asia are about seven kinds of tree shrews (Fig. 68A), in which the fingers have claws instead of nails. The skull and teeth are extremely generalized, like those of primitive insectivores, and, in fact, the tree shrews have been placed by some recent workers in the primates and by others in the Insectivora. In any case, they appear to be the surviving members of a stem group representing the ancestry of primates.

A serious difficulty in understanding the classification of primates has been the excessive accumulation of unnecessary taxonomic names, especially of species, genera, and families. These names have to be considered and disposed of by those who work in the field, but they are very largely due to the zeal of some students of man who were interested in man's relatives but did not know zoology or the meaning of taxonomic groups and who overemphasized trivial variations in describing newly found, often fragmentary, specimens. For example, at least seventeen generic names have been applied to animals that are now placed in the primitive ape genus *Dryopithecus* (Miocene and Pliocene of the eastern hemisphere). Nine or more generic names have been used for prehistoric men (not including *Australopithecus*) that belong in *Homo,* and most of these were within one or two species. These names impede instead of helping our understanding of human origins, as well as of the order as a whole, and they have to be dropped into synonymy (discarded), along with innumerable other useless names for almost every valid genus of primates, before the picture becomes clear. This problem is nearly solved.

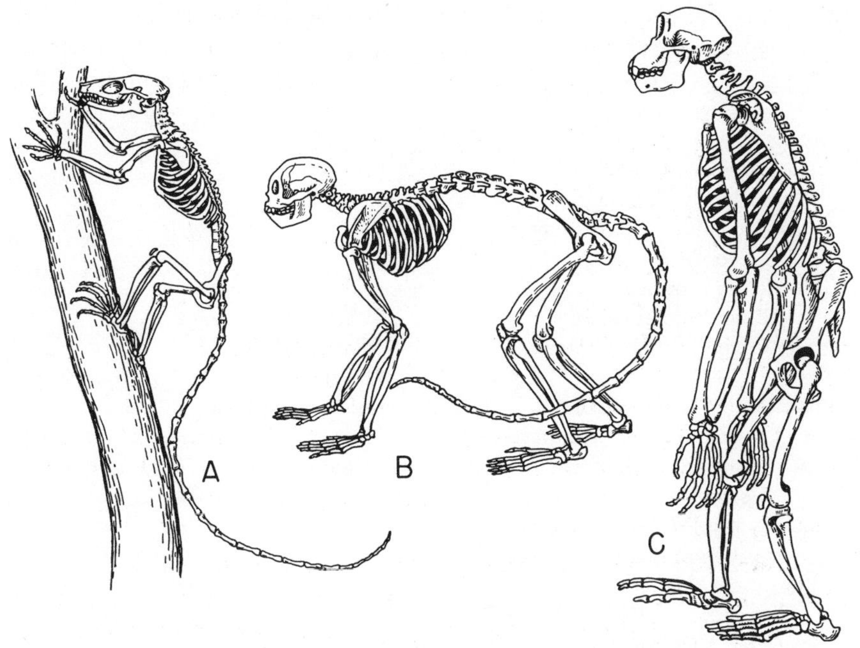

Figure 68. Skeletons of primates: (A) *Ptilocercus*, pen-tailed tree shrew; (B) monkey, *Cercopithecus;* (C) chimpanzee, *Pan.* (Not to scale.) (Redrawn by permission of Quadrangle Books from *The Antecedents of Man* by W. E. LeGros Clark, copyright © 1959 by Edinburgh University Press.)

A less serious difficulty concerns reasonable differences of opinion about classification, which are largely matters of judgment on the scope and the rank of the larger categories of primates. This will eventually be settled among zoologists as the anatomical and fossil evidence becomes more complete; the primates are not common or widespread fossils, and their study has not come as near to full understanding as that of some other groups. It may be useful here to give in summary form not one, but two, recent classifications in order to show the kind of differences between them. The rank (for example, suborder, infraorder, or superfamily) is not so important as the grouping, in this case of families (omitting genera and species). The succession of groups as listed is not, of course, an evolutionary succession, because there has been branching of the tree and not merely a single line.

Except as noted, the column on the left is based on Simpson (1945), although it ought not to be taken as indicating his current views. That on the right is based on Romer (1966). (An asterisk indicates that the family is extinct.)

Order Primates

Suborder Prosimii Infraorder Lemuriformes Superfamily Tupaiioidea *Anagalidae, Tupaiidae (tree shrews)	
(Uncertain infraorder *Phenacolemuridae, *Carpolestidae)	_Suborder Plesiadapoidea_ *Phenacolemuridae, *Carpolestidae, *Plesiadapidae
Superfamily Lemuroidea *Plesiadapidae, *Adapidae, Lemuridae, Indridae (lemurs) Superfamily Daubentonioidea Daubentoniidae (aye-aye) Infraorder Lorisiformes Lorisidae (lorises, galagoes)	_Suborder Lemuroidea_ *Adapidae, Lemuridae, Lorisidae, Daubentoniidae (lemurs, lorises, galagoes, aye-aye)
Infraorder Tarsiiformes *Anaptomorphidae, Tarsiidae (tarsiers)	_Suborder Tarsioidei_ *Anaptomorphidae, *Omomyidae, Tarsiidae (tarsiers), *Microsyopidae
Suborder Anthropoidea Superfamily Ceboidea Cebidae, Callithricidae (New World Monkeys)	_Suborder Platyrrhini_ Cebidae, Callithricidae (New World monkeys)
Superfamily Cercopithecoidea Cercopithecidae (Old World monkeys)	_Suborder Catarrhini_ Superfamily Cercopithecoidea Cercopithecidae (Old World monkeys) Superfamily Parapithecoidea *Parapithecidae
Superfamily Hominoidea *Parapithecidae, *Oreopithecidae (Simpson, 1963), Pongidae (apes), Hominidae (man)	Superfamily Hominoidea *Oreopithecidae, Pongidae (apes), Hominidae (man)

Among the tree shrews, considered primates by Simpson but not (later) by Romer, the only fossil genus is *Anagale,* described by Simpson in 1931 from the lower Oligocene of Mongolia. The dental formula was $\frac{3\ 1\ 4\ 3}{3\ 1\ 4\ 3}$, characteristic of the most primitive placentals. (This means that on either the right or the left side of the upper jaw and also of the lower jaw there are 3 incisor teeth, 1 canine, 4 premolars, and 3 molars; the total is

44.) The orbit was not closed behind and the digits of the hind foot bore nails instead of claws. The other family, Tupaiidae, is not known from fossils but from six living genera in southeastern Asia, and as far east as Borneo and the Philippines. The genus *Tupaia* and four others are diurnal, arboreal, more or less omnivorous creatures feeding mainly on insects. *Ptilocercus*, the pen-tailed tree shrew (Fig. 68A), is nocturnal and somewhat less specialized in its dentition. In the Tupaiidae the dental formula is $\frac{2\ 1\ 3\ 3}{3\ 1\ 3\ 3}$, and the lower incisors are specialized for use as a comb or scratcher. They are nearly horizontal in position, slender, and close together. The digits carry claws instead of nails. The specialization of the incisors was present also in *Anagale* and is carried even further in most of the lemurs, so that at first sight it would appear to be a character of primitive primates in general, but in fact there is no need for this assumption, as we shall see.

Lemurs in the broad sense include a number of families, arranged differently in the two lists. The peculiar genus *Daubentonia* is the nocturnal aye-aye, living in the tropical forests of Madagascar. All the digits have claws except for a flat nail on the hallux. The middle digit of the hand is extremely slender and is used as a probe for reaching into holes in wood to extract grubs. The incisor teeth are highly modified in a manner analogous to those of rodents, being chisel-shaped, heavy, and with deep roots. There is a diastema (a gap) between them and the cheek teeth. The dental formula is $\frac{1\ 0\ 1\ 3}{1\ 0\ 0\ 3}$. *Daubentonia* is thought to have been derived from among the earlier lemurs, but its exact source is not known.

Certain Paleocene and Eocene "lemuroids," as well as some Eocene tarsioids, showed a reduction of cheek teeth, development of a diastema, and rodent-like incisors that are comparable to those of *Daubentonia*. The relationships among lemuroids that show this specialization are not yet settled, but the prevalent opinion (Clark, 1959) is that it developed independently in several groups.

In almost all Recent lemurs the dental formula is $\frac{2\ 1\ 3\ 3}{2\ 1\ 3\ 3}$ or $\frac{2\ 1\ 2\ 3}{2\ 0\ 2\ 3}$. These animals possess comblike incisors, to which the slender, horizontal canines of the lower jaw are added, but the upper canines are generally large and prominent. The digits have nails, except for the second one on the hind foot, which bears claws used for scratching. These animals have no fossil record prior to the Pleistocene. They vary from squirrel-like or somewhat monkey-like forms to the giant *Megaladapis*, which was perhaps the largest of all primates.

Lorises are known from the Pliocene to Recent in Asia and Recent in Africa. *Loris* and *Nycticebus* are the lorises of southeastern Asia. *Arcto-*

cebus and *Perodicticus* are the pottos, and *Galago* and *Euoticus* are bush babies or galagoes of Africa. In the colloquial sense, all these are included in the lemurs. (The so-called "flying lemurs" of southeastern Asia are not primates but are placed in an order Dermoptera, which has apparently been separate from insectivores and primates since the Paleocene. As is customary, we should note that they are not lemurs and do not fly but glide in a manner analogous to the flying squirrel.)

The tarsioids include two families, one extinct since the Lower Oligocene, and the other containing only the living *Tarsius* of the East Indies. The dental formula in *Tarsius* is $\frac{2\ 1\ 3\ 3}{1\ 1\ 3\ 3}$. There is almost no horizontal projection of the anterior teeth, and the canines of both jaws are well developed. Clark (1959, p. 110) says:

> So far as it is possible to conjecture the origin of the tarsioids on the basis of the dentition of the known genera, it is reasonable to infer that the common ancestral stock was characterized by the generalized placental dental formula, and by quite generalized features of all the component elements of the dentition. In other words, it seems likely that the dentition of the ancestral stock would have shown no distinctive characters by which it could be labeled lemuroid. It has commonly been assumed that in the evolutionary sense tarsioids arose from lemuroids. If this were so, separation of the one from the other must have occurred at a stage when the dentition was still exceedingly primitive and generalized.

On the other hand, *Anaptomorphus,* an Eocene tarsioid, had lost both the first and second premolar of the lower jaw, those of the upper jaw having been lost already, and therefore had acquired the dental formula $\frac{2\ 1\ 2\ 3}{2\ 1\ 2\ 3}$, as is present in the Old World monkeys, apes, and man.

Anthropoidea, often used as a suborder of the primates, are not with complete certainty a natural group and may not be maintained as a suborder. This is because of the widely separated and perhaps independent ceboid monkeys of the New World, the source of which may have been different from that of the other "anthropoids," in the Old World. Concerning ceboids Simpson (1945, p. 185) says: "As far as the present record goes ceboids thus appeared suddenly and in almost modern guise in the mid-Tertiary of South America. The most reasonable hypothesis is that ceboids arose from one of the Paleocene or Eocene prosimian stocks of North America and that their early deployment or indeed almost all their history occurred in the more tropical parts of South America, where Tertiary fossils are extremely rare. This is however only a hypothesis." The dental formula is in almost all cases $\frac{2\ 1\ 3\ 3}{2\ 1\ 3\ 3}$, and the ceboids must be considered a natural group that is both anatomically and geographically rather restricted. Some of them, the Cebidae, show form and habits closely parallel to those of the typical arboreal monkeys of Africa and Asia, and among cebids a few have gone farther in developing a prehensile tail.

The cercopithecoids (Old World monkeys) and hominoids (apes and man) form a natural group with several features in common, including dentition, $\frac{2\ 1\ 2\ 3}{2\ 1\ 2\ 3}$, and a narrow nasal septum. Their history was entirely in the continents of Asia, Africa, and Europe until, in the last few thousands of years, man reached Australia and the Americas. In the monkeys (Fig. 68B) there is a diversity of adaptations, the most highly arboreal being such as the long-legged and long-tailed langurs of the Orient and *Colobus* of Africa. The macaques (*Macaca*) are at home in trees or on the ground; one of them, the Barbary ape, is strictly terrestrial and has no tail. Most members of the genus are Asiatic, such as the familiar rhesus monkey. Baboons are primarily terrestrial members of the same family.

Although the Conference on Classification and Human Evolution held in 1962 (papers published in 1963) reflected a diversity of opinion among those who participated, it accomplished a great deal in agreeing upon a simplified classification of Hominoidea. Three families are included: Pongidae, Oreopithecidae, and Hominidae.

Apidium of the early Oligocene of Egypt and *Oreopithecus*, dating from approximately the beginning of the Pliocene in southern Europe, have characters of the lower molars that are considered more primitive than those of any other ape or man. They lack the diastema (gap behind the canines) found in Pongidae and in that respect agree with Hominidae. In the skeletal characters there is a resemblance to man in the proportions of the limbs, the short face, presence of a large heel, and several other minor characters. Thus it seems that *Oreopithecus* represents a branch from a stock that may have led to man, but it is not, itself, directly on the line of human ancestry. Although in 1945 Simpson placed *Oreopithecus* and *Apidium* doubtfully among Old World monkeys, he later (1963) concluded that they should be a family in the Hominoidea; this paper, incidentally, on the meaning of taxonomic statements, should be required reading for anyone working on primates.

Several genera of apes are known from the Oligocene, Miocene, and Pliocene. Most of them represent early stages in divergence of dentition and other characters from the hominid line, but the recently (1966) discovered Oligocene genus *Aegyptopithecus* is thought to be a satisfactory common ancestor for both lines, the modern apes and man. The divergence of later apes was based largely, according to Washburn (1963), on locomotor adaptations, especially use of the arms in brachiation (swinging), but Washburn was careful to point out that brachiation is not the only means of locomotion in apes. They are in some circumstances quadrupedal or even bipedal and resort to brachiation when it is speedy and convenient for progress among the branches of trees. The gibbons reach the extreme of specialization in the brachiating mechanism, but the other larger apes, *Pongo* (orang) and *Pan* (gorilla, chimpanzee), have become more specialized in the structure of the head, especially in enlargement of the jaws and canine teeth (Fig. 68C). Development of prognathism (prominent jaws) and of a strong supraorbital crest is parallel to but independent of the same characters in species of *Australopithecus* and *Homo*.

So far there is no fossil record of true Hominidae prior to the late Pliocene, unless *Ramapithecus*, known only from teeth, is a hominid. This is *not* evidence that the distinctive characters of Hominidae evolved rapidly from those of an advanced pongid rather than reaching back into the Miocene or earlier. It is simply a lack of evidence one way or the other. But the skeletal and limb proportions obviously did reach the "human" stage long before those of skull and brain. It is important to notice that the size of the brain of modern man is not to be taken as a character of the family Hominidae as a whole, nor are associated features such as the position of the foramen magnum, the fairly vertical face, the presence of a chin, and the shortness of the tooth row. All these can be seen to have evolved within the genus *Homo* itself during the Pleistocene. The human skeleton and limbs are specialized for walking and running in the open, although man is quite capable of brachiation or even walking as a quadruped in some circumstances. The shift of emphasis is relative rather than absolute.

Living in southern and eastern Africa and perhaps more widely still during the Pleistocene was a primitive genus of men, *Australopithecus* (Fig. 69). The earliest well-confirmed date, 1,750,000 years ago, is for

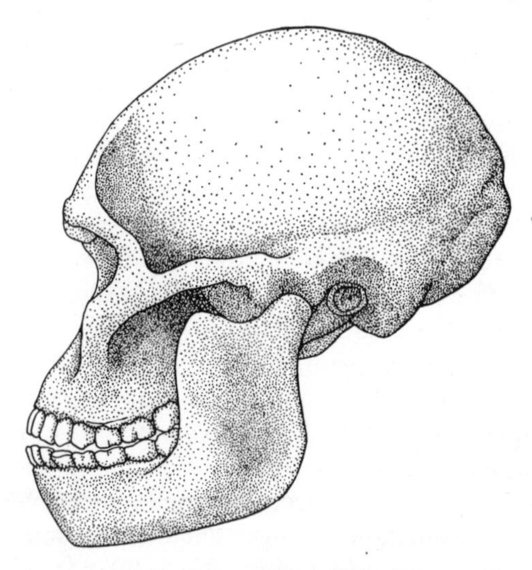

Figure 69. Skull of *Australopithecus africanus,* adult.

those discovered in the lower bed of Olduvai Gorge in Tanzania and described by Leakey under the names *Zinjanthropus boisei* and *Homo habilis*. The characteristics of these do not differ, according to a series of papers by Robinson (1964, 1965) from those of *Australopithecus* in South Africa sufficiently to justify assignment to another genus or even a species different from *A. africanus*. See also Clark (1964). Further material from a higher level in the gorge belongs to *Homo erectus*. *Australopithecus* from South Africa has been well studied, especially by Broom and Robinson, who recognize a genus *Paranthropus* in addition, but this does not seem sufficiently different to most others working in the same field. There are a number of skeletons of both sexes, adults and children, from several different localities, principally in caves and fissure deposits of limestone. Their age is considered on the basis of the associated extinct mammals to be early Pleistocene, perhaps as late as the first interglacial stage in Europe, but this is not more than half as old as the lower levels of Olduvai Gorge. We must either extend our concept of the length of the Pleistocene to some 2 or 3 million years (as some are doing) or place the older finds in the Pliocene.

The information so far gained from this assortment of early human remains indicates that the entire skeleton with the exception of the skull was closely similar to that of modern man, *Homo sapiens*, and many of the bones are indistinguishable. Therefore there can be no doubt that *Australopithecus* walked and ran just as men do today. The mammalian fauna associated with *Australopithecus* indicates a relatively dry and open, rather than completely forested, environment, indeed much like that persisting in southern and eastern Africa today, but the skulls show a startling contrast to those of modern man. In adults the face was large, sloping, and the jaws enormous. The bridge of the nose did not protrude, but massive ridges surmounted the eyes. The teeth, although distinctly human in shape and character, were large, befitting the prominent upper and lower jaws. There was no chin. Above the level of the eyebrow ridges there was no forehead in the usual sense of the word but at most a gentle upward slope of the roof of the cranium to a maximum not much higher than the level of the supraorbital ridges. Posteriorly the cranium curved gently down to a moderately strong horizontal occipital ridge marking the edge of the attachment of muscles from the back of the neck. The position of the condyles and the foramen magnum was slightly behind the middle of the cranium, that is, behind the position in which they are located in modern man. But perhaps the most distinctive feature of the skull in contrast to that of *Homo sapiens* was the relatively small cranium and its elevated position relative to the face and jaws. It is as if the brain and its enclosing case could not keep up in growth rate with the face. The cranial capacity varied from about 450 to 600 or possibly at most somewhat over 800 cubic centimeters, in contrast with a minimum of about 1200 in normal adult *H. sapiens*. But the inner surface of the brain case gives indication of abundant convolutions of the cortex. These features show that in the early stages of human evolution the locomotor and dental structure had evolved to an essentially modern condition long before the structure of the skull and the size of the brain. In the latter feature, *Australopithecus* was in about the same stage as has been attained independently in the largest modern apes. In fact, it is important to remember that the apelike appearance of the face and skull of some early men was the result of an independent evolution of certain features of the jaws and face, separate from that in modern apes, perhaps associated with an increase in size in both groups.

Just as the present view of *Australopithecus* is that it gathers together in one genus several closely similar ancient men which had been described under different generic names but which may belong to one or perhaps two or three species, so *Homo* now contains not only recent man but what were formerly considered to be several genera and species of Pleistocene man. With the increasing discoveries of recent years, it has become apparent that the differences among these "kinds" of men are not greater than those among species within a single genus elsewhere among mammals. Two species of *Homo* are presently distinguished. The older one with smaller brain, *H. erectus* (Fig. 70), includes the so-called Java ape-man, originally named *Pithecanthropus erectus;* the later discovery, *P. robustus;* the Peking man, *Sinanthropus pekingensis* (Fig. 71); an Algerian fossil named *Atlanthropus;* a partial skull from Olduvai Gorge reported by Leakey (1961) with a potassium–argon date of 490,000 years; and possibly the Heidelberg jaw from Germany, estimated to have an age of 400,000 years. *Gigantopithecus* from southeastern Asia may be omitted from this list. Clark (1964) thinks it may have been a giant ape in which the teeth were less strongly specialized than in the modern large apes.

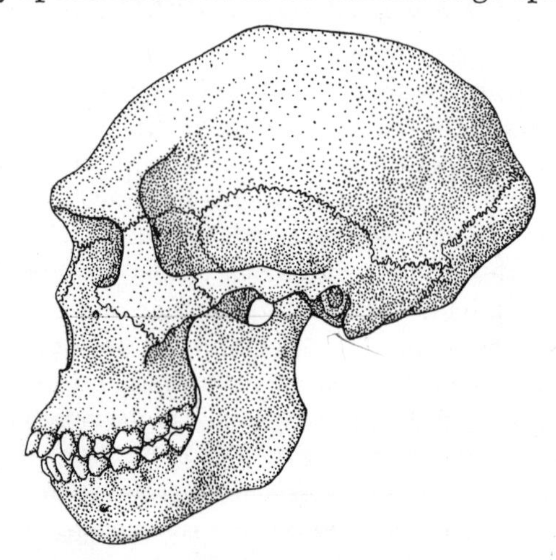

Figure 70. Skull of *Homo erectus* (Java)

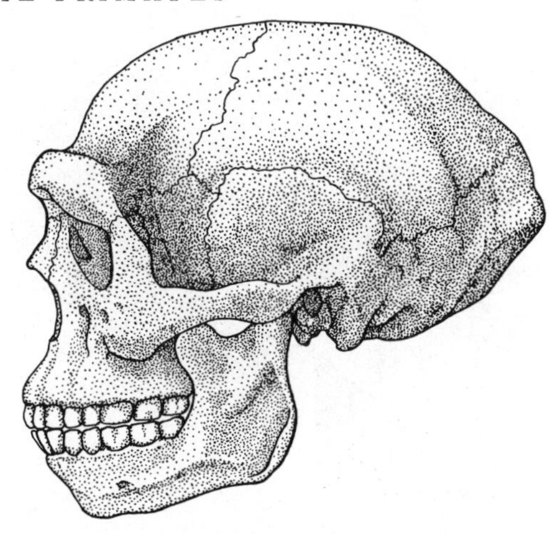

Figure 71. Skull of *Homo erectus* (Peking).

The remains of *H. erectus* found in Java are of early Pleistocene age, but those of China and Algeria are probably not older than middle Pleistocene, that is, the second interglacial period. The cranial cavity had a mean capacity of about 1000 cubic centimeters, those of Java ranging between 775 and 900, but four specimens from Peking varied from 850 to 1300, with a mean of 1075. Thus the known range of variation attained the lower limits of that of *H. sapiens* but in general was much less. The cranial wall was thick. There were heavy supraorbital ridges, but the forehead was sloping and little elevated. The mandible, although large and heavy, did not reach the proportions of that of *Australopithecus*. It receded at the chin. The teeth were relatively large, with canines that sometimes projected slightly and interlocked; there was a small upper diastema (gap). The limbs and stature of the body were as in *H. sapiens*. It is not impossible that more than one species of man may be represented among the incomplete fossils of *H. erectus* that have been found, but there is no clear evidence that this is the case.

The species *H. sapiens* (Figs. 72 to 74) is now thought to include not only Recent and Cro-Magnon man, but the much older skulls of Swanscombe, Steinheim, Kansano, and Ehringsdorf; the skeletons of Mt. Carmel; and others. Among these are the so-called "generalized Neanderthal" type, as seen in the population at Mt. Carmel. The later, more extreme, "classical" Neanderthal type ranged widely in Europe and North Africa during the early part of the last (Würm) glaciation, and then became extinct. There is still some difference of opinion as to whether it should be classified as a separate species, *H. neanderthalensis*. Clark (1964) favors doing so, solely on the degree of divergence from *H. sapiens*, which he says can be recognized even in juvenile skulls. In view of the lack of clear-cut evidence concerning the distribution and evolutionary stages of the popu-

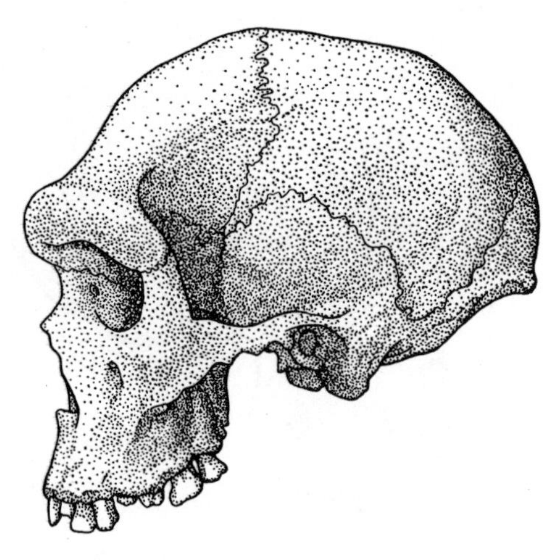

Figure 72. Skull of *Homo sapiens* (Rhodesian man).

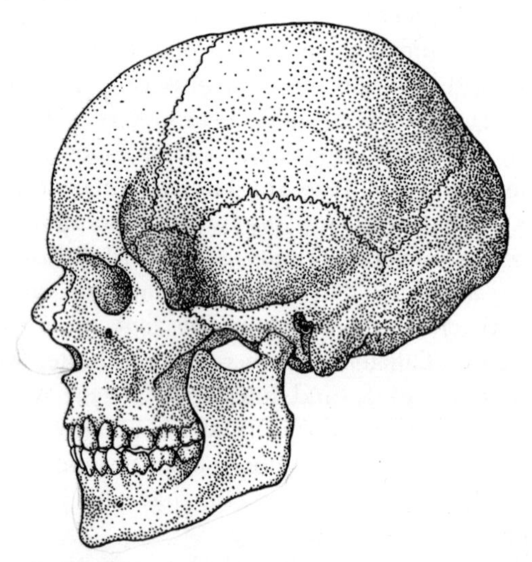

Figure 73. Skull of *Homo sapiens* (Cro-Magnon).

lations of Neanderthal man, I cannot see that any harm is done by including even the extreme Neanderthal as a subspecies (geographical race) of *H. sapiens.*

Homo sapiens must have originated at least 200,000 years ago and its ancestry almost certainly lay within some population of *H. erectus. Sapiens* is a large-brained species in which the cranial walls are usually thin, the forehead higher and steeper than in *erectus,* the teeth smaller, the face more nearly vertical, the eyebrow ridges usually less prominent, but in early representatives of *H. sapiens* the supraorbital ridges were distinct

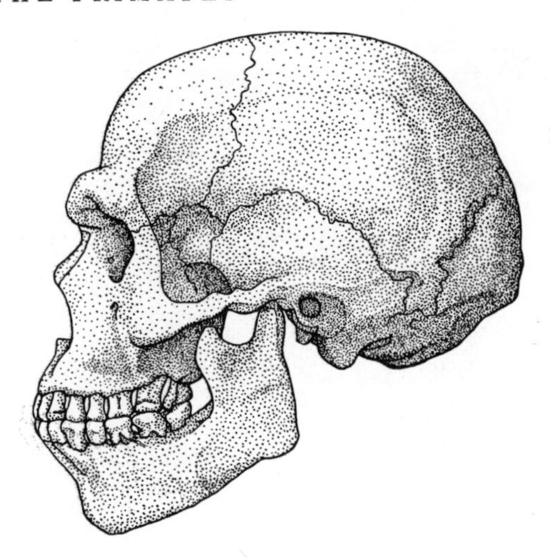

Figure 74. Skull of *Homo sapiens* (early Neanderthal).

and became again relatively big in the Neanderthal race. The forehead was not so high nor the chin so prominent in early *H. sapiens.*

Skulls of *H. sapiens* were found in a cave at Fontéchèvade in France in 1947 under an undisturbed stalagmite layer, which in turn was covered by deposits containing evidence of Mousterian (Neanderthal) culture. These early skulls are associated with a warm-temperate fauna and date from the third interglacial period. The cave inhabitants of Mt. Carmel in Palestine are thought to have existed at least 10,000 years and to represent shifting populations of the last interglacial and early part of the Würm glacial period. The skulls have strong supraorbital ridges but are otherwise quite variable. The earlier individuals may represent a transition toward the classical Neanderthal type, but the variability is not thought to indicate interbreeding.

As far as we know, then, there has been an essentially modern type of man present in the world since the second interglacial period but with divergence or variation in the direction of the Neanderthal type also occurring, up to the early phases of the last glaciation. Following the disappearance of the Neanderthal "race," the known populations and cultures have been those of modern men, that is, men whose characteristics come within the range of variation seen among human beings today.

The table below summarizes certain time relationships of early man in the Pleistocene. It is only an approximation. Corresponding names are given for European and American glacial and interglacial stages, but man did not arrive in the New World until near the close of the Wisconsin glaciation, and his prior history is therefore limited to the eastern hemisphere.

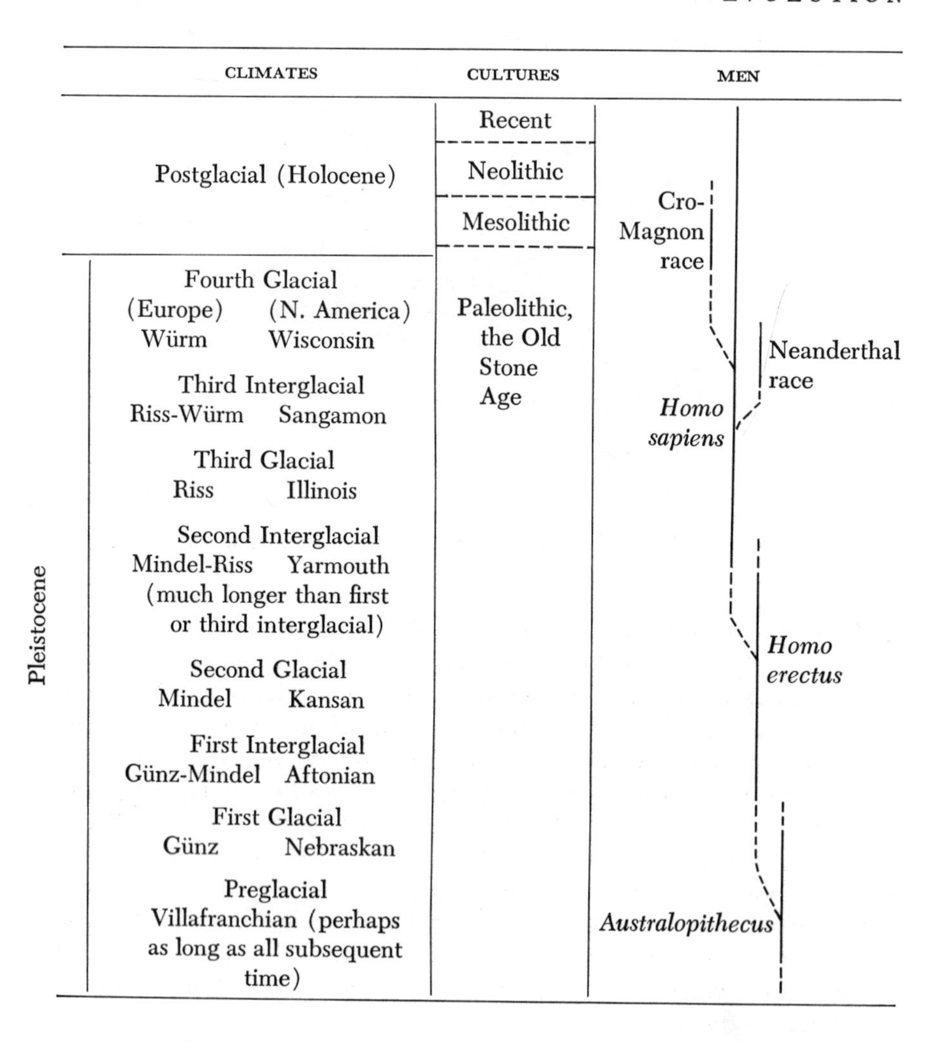

REFERENCES

Clark, W. E. 1959. *The Antecedents of Man.* Edinburgh Univ. Press, Edinburgh; also Quadrangle Books, Chicago, 1960.

———. 1964. *The Fossil Evidence for Human Evolution,* 2nd ed. Univ. Chicago Press, Chicago.

Gregory, W. K. 1920. On the structure and relations of *Notharctus,* an American Eocene primate. *Mem. Am. Mus. Nat. Hist.* 3(2): 49–243.

Leakey, L. S. B. 1961. New finds at Olduvai Gorge. *Nature 189:* 649–650.

Robinson, J. T. 1964. Some critical phases in the evolution of man. *Archaeol. Bull.* 19(73): 3–12.

———. 1965. *Homo "habilis"* and the Australopithecines. *Nature 205:* 121–124.

Romer, A. S. 1966. *Vertebrate Paleontology,* 3rd ed. Univ. Chicago Press, Chicago.

Simpson, G. G. 1945. The principles of classification and a classification of mammals. *Bull. Am. Mus. Nat. Hist. 85:* 1–350.

———. 1963. The meaning of taxonomic statements, pp. 1–31. In S. L. Washburn (ed.), *Classification and Human Evolution.* Aldine, Chicago.

Washburn, S. L. 1963. Behavior and human evolution. In S. L. Washburn (ed.), *Classification and Human Evolution.* Aldine, Chicago.

INDEX